Metal and Ceramic Biomaterials

Volume I
Structure

Editors

Paul Ducheyne, Ph.D.
Associate Professor of
Biomedical Engineering
Associate Professor of
Orthopaedic Surgery Research
University of Pennsylvania
Philadelphia, Pennsylvania

Garth W. Hastings, Ph.D., D.Sc., C.Chem., F.R.S.C., F.R.R.I.
Acting Head
Biomedical Engineering Unit
North Staffordshire Polytechnic
Honorary Scientific Officer
North Staffordshire Health Authority
Medical Institute
Hartshill, Stoke-on-Trent
England

CRC Series in Structure-Property Relationship of Biomaterials
Series Editors-in-Chief
Garth W. Hastings and Paul Ducheyne

CRC Press
Taylor & Francis Group
Boca Raton London New York

CRC Press is an imprint of the
Taylor & Francis Group, an **informa** business

First published 1984 by CRC Press
Taylor & Francis Group
6000 Broken Sound Parkway NW, Suite 300
Boca Raton, FL 33487-2742

Reissued 2018 by CRC Press

Library of Congress Cataloging in Publication
Main entry under title:

Metal and ceramic biomaterials.

(CRC series in structure-property relationship of biomaterials)
Bibliography: p.
Includes index.
Contents: v. 1. Structure — v. 2. Strength and surface related behavior.
1. Ceramics in medicine. 2. Metals in surgery.
I. Ducheyne, Paul. II. Hastings, Garth W. III. Series.
R857.C4M47 1984 610'.28 83-15018
ISBN 0-8493-626I-X (v. 1)
ISBN 0-8493-6262-8 (v. 2)

A Library of Congress record exists under LC control number: 83015018

Publisher's Note
The publisher has gone to great lengths to ensure the quality of this reprint but points out that some imperfections in the original copies may be apparent.

ISBN 13: 978-1-315-89533-8 (hbk)
ISBN 13: 978-1-351-07443-8 (ebk)

Visit the Taylor & Francis Web site at http://www.taylorandfrancis.com and the
CRC Press Web site at http://www.crcpress.com

SERIES PREFACE

Biomaterials science is concerned with surgical implants and medical devices and their interaction with the tissues they contact. Their study, therefore, includes not only the properties of the materials from which they are made, but also those of the tissues which will accept them. Metals, ceramics, and macromolecules are the artifacts. Bone tendons, skin, nerves, and muscles are among the tissues studied. Prosthetic materials, implants, dental materials, dressings, extra corporeal devices, encapsulants, and orthoses are included among the applications.

It is not only the materials *per se* which interest the biomaterials scientist, but also the interactions in vivo, because it is at the interface between implant and tissues that the success of a procedure will be decided. This approach has led to the concept of a more aggressive role for biomaterials in the actual treatment of disease. Macromolecular drug delivery systems are receiving considerable attention, especially those with the capacity for targeting specific sites in the body. Sensing and control of body processes is a logical extension of this. There is much to be done before these newer developments become established.

The science of biomaterials has grown and developed over the last few years to become an accepted discipline of study. It is opportune, therefore, to systematize the study of biomaterials in order to improve their application in medical science, since that is the end point of all studies. That is the aim of this series of books on *Structure-Property Relationships in Biomaterials*. Knowledge of structure and the influence on properties is fundamental to any materials science study; it is a more complex problem to obtain the knowledge from tissue materials, as the living organism has a great capacity for change and adaptation in response to a stimulus. The stimulus may be chemical, electrical, or mechanical. The biomaterials scientist endeavors to identify and to use these stimuli and responses to improve the in vivo acceptability of the materials.

Many institutions and agencies have promoted the science of biomaterials. Societies now exist for this purpose. The Biological Engineering Society (U.K.) founded in 1960 formed a Biomaterials Group in 1974. In the same year the Society for Biomaterials was founded in the U.S. The European Society for Biomaterials (1976) was followed by Canadian and Japanese Societies (1979). All societies play a major role in disseminating knowledge through conferences and publications.

This series is complementary to these society activities. It is hoped that it will not only provide a basis of knowledge, but also its own stimulus for further progress. The series is inevitably selective. In part this is due to the editors' choice, in part to the availability of authors. The editors wish to thank those who fulfilled their agreements. Without them this series would not have been possible.

G. W. Hastings
Series-Editor-in-Chief

VOLUME PREFACE

The understanding of the in vivo performance of synthetic materials is largely dependent upon a profound knowledge of the properties of the materials in question. Analogous to materials science in its broadest sense, the basis for biomaterials science is formed by microstructural theory. It is, therefore, that in this series on structure-property relationships in biomaterials a substantial part is devoted to the analysis of the basic properties of the various synthetic biomaterials. In addition, the effect of microstructural aspects on properties is considered at great length.

The study of metallic and ceramic biomaterials is intimately interlinked because the microstructural aspects and the research methodologies are founded on the same basis. This is demonstrated first in the chapter of Dr. Heimke, who discusses the structure of metals and ceramics, and second, in the chapter of Drs. Arkens and Ducheyne who analyze some of the more recent surface analytical techniques. Those techniques have been equally employed for metals and ceramics. Two subsequent chapters deal with the influence actual manufacturing has on both the microstructure and the properties of either metals (authored by Dr. Pilliar) or ceramics (written by Dr. Doremus).

The second volume of *Metal and Ceramic Biomaterials* builds upon the foundation laid in the first volume by analyzing various properties and by discussing these in light of the microstructural theory outlined in the first volume. Strength related behavior is treated first. Dr. Semlitsch surveys the mechanical properties of implant metals used for artificial hip joints. Drs. Soltesz and Richter describe the mechanical behavior of various bioceramic materials. A third chapter relating mechanical behavior to microstructural detail deals with the shape memory alloys which have considerable potential as biomaterials. In addition to the description of strength aspects, Dr. Kousbroek summarizes the various biocompatibility studies on these metals. These studies are, quite naturally, related to the surface properties which are presented in the second part of Volume II. Drs. Lycett and Hughes survey the corrosion behavior of metals. Drs. Van Raemdonck, Ducheyne, and De Meester analyze the bioreactivity of a typical class of ceramics, viz. calcium phosphate. An effort is made to relate the bioreactivity to the microstructural chemical and physical detail. Drs. Dumbleton and Higham analyze the field of surface coatings and point to the great potential for the field of biomaterials of many existing technologies.

The editors have enjoyed collaborating with the contributors to these volumes. We owe great appreciation and gratitude to those who eventually made it possible to have these books produced. Without each single contributor it would have been hard, if not impossible to present the information on some of the more important properties of biomaterials. Due thanks are also expressed to Rita De Laet who diligently took care of the secretarial work.

Last but not least, thanks go to our families. We may think that without their delightful distraction we could have finished these books sooner. However, they rightfully claim that without them we would never have had the perseverance needed to edit this series.

Paul Ducheyne
Garth Hastings

THE EDITORS

Garth W. Hastings, D.Sc., Ph.D., C.Chem., F.R.S.C., is a graduate of the University of Birmingham, England with a B.Sc. in Chemistry (1953) and a Ph.D. (1956) for a thesis on ultrasonic degradation of polymers. After working for the Ministry of Aviation he became Senior Lecturer in Polymer Science at the University of New South Wales, Sydney, Australia (1961 to 1972). During this time he was Visiting Professor at Twente Technological University, Enschede, The Netherlands (1968-69), advising on their program in biomedical engineering. While in Australia, he became associated with Bernard Bloch, F.R.C.S., Orthopedic Surgeon, Sydney Hospital, and began a fruitful collaboration in the uses of plastics materials in surgery.

In 1972 he returned to England as Principal Lecturer in the Biomedical Engineering Unit of the North Staffordshire Polytechnic and the (now) North Staffordshire Health District with responsibility for research. With a particular interest in biomaterials research his own work has encompassed carbon fiber composites for surgical implants, adhesives, bioceramics, prosthesis performance in vivo, and electrical phenomena in bone. He is a member of British and International Standards Committees dealing with surgical implants and of other professional and scientific bodies, including Companion Fellow of the British Orthopaedic Association and Editor of the international Journal *Biomaterials*. He was elected President of the Biological Engineering Society in the U.K. (B.E.S.) in October, 1982. He was awarded a D.Sc. from the University of Birmingham in 1980 for a thesis in the field of biomedical applications of polymers. He has recently been appointed Acting Head of the department.

Paul Ducheyne, Ph.D. obtained the degree of metallurgical engineering from the Katholieke Universiteit Leuven, Belgium, in 1972. Subsequently he worked at the same university towards a Ph.D. on the thesis "Metallic Orthopaedic Implants with a Porous Coating" (1976). He stayed one year at the University of Florida as an International Postdoctoral N.I.H. Fellow and a CRB Honorary Fellow of the Belgian-American Educational Foundation. Thereafter he returned to the Katholieke Universiteit Leuven. There he was a lecturer and a research associate, affiliated with the National Foundation for Scientific Research of Belgium (NFWO). He recently joined the University of Pennsylvania, Philadelphia, as an Associate Professor of Biomedical Engineering and Orthopedic Surgery Research.

Dr. Ducheyne has published in major international journals on mechanical properties and design of prostheses, porous materials, bioglass, hydroxyapatite, and microstructural methods of analysis of biomedical materials. He is member of the editorial board of *Biomaterials, Journal of the Engineering Alumni of the University of Leuven, Journal Biomedical Materials Research,* and *Journal Biomechanics and Comtex System for Biomechanics and Bioengineering.*

He became active in various societies and institutions and has held or is holding the positions of Chairman-Founder of the "Biomedical Engineering and Health Care Group" of the Belgian Engineering Society, Secretary of the European Society for Biomaterials and member of the Board of Directors of Meditek (Belgian Institution to promote biomedical industrial activity).

CONTRIBUTORS

Oger F.. Arkens, Ph.D.
First Assistant
Department Metaalkunde
Katholieke Universiteit
Leuven, Belgium

P. De Meester, Ph.D.
President of Exact Sciences Group
Department Metaalkunde
Katholieke Universiteit
Leuven, Belgium

Robert H. Doremus, Ph.D.
Professor of Glass and Ceramics
Materials Engineering Department
Rensselaer Polytechnic Institute
Troy, New York

John H. Dumbleton, Ph.D.
Howmedica, Inc.
Rutherford, New Jersey

G. Heimke, Ph.D.
Friedrichsfeld GmbH
Mannheim, West Germany

Paul Higham, Ph.D.
Howmedica, Inc.
Rutherford, New Jersey

A. N. Hughes, Ph.D.
Principal Scientific Officer
A.W.R.E.
Reading, England

Ronald Kousbroek, M.Sc.
Delfzijl
The Netherlands

R. W. Lycett, Ph.D.
Ministry of Defense
Whitehall
London, England

Robert M. Pilliar, Ph.D.
Professor
Faculty of Dentistry
Department of Metallurgy and Materials Science
Toronto, Ontario
Canada

H. Richter, Ph.D.
Senior Scientist
Fraunhofer-Institut für Werkstoffmechanik
Freiburg, W. Germany

M. Semlitsch, Ph.D.
Department of Research and Development
Winterthur, Switzerland

U. Soltész, Ph.D.
Senior Scientist
Fraunhofer-Institut für Werkstoffmechanik
Freiburg, W. Germany

W. Van Raemdonck, M.Sc.
Department Metaalkunde
Katholieke Universiteit
Leuven, Belgium

TABLE OF CONTENTS

Volume I

Volume II

Chapter 1

THE STRUCTURE, PROPERTIES, AND FUNCTIONAL BEHAVIOR OF BIOMATERIALS

P. Ducheyne and G. W. Hastings

TABLE OF CONTENTS

I. THE SCIENCE OF BIOMATERIALS

The science of biomaterials has previously been defined as the science dealing with the development, evaluation, and application of special substances to meet specifications for research and practice in biology and medicine, with particular reference to materials which come into contact with tissue. In order to be more explicit than this general definition, it is useful to consider the description of bioengineering, formulated by the National Academy of Engineering, Washington, D.C., in 1971:

1. The *application* of bioengineering concepts and technology to scientific inquiries into biological phenomena as a basis for advancing the understanding of biological systems and medical practices
2. The *utilization* of engineering concepts and technology in the development of instrumentation, materials, diagnostic and therapeutic devices, artificial organs, and other constructs relevant to applications in biology and medicine
3. The application of engineering concepts, methodology, and technology to the improvement of health service delivery systems in the broad context of interrelated institutions (hospitals, clinics, governmental units, universities, industries, etc.) as well as within the specific confines of individual components of the health care system

In its broadest sense, the science of biomaterials studies only biological materials (as described in the listing above) (1) and manmade materials (2) and is clearly not involved in the activities described in (3).

As an example we may consider the medical discipline of orthopedic surgery. An important area of study is the development and improvement of temporary and permanent implants made of metals, polymers, or ceramics. This study relates to the utilization of technology. A further area of intensive research is concerned with biological materials. The mechanical and elastic properties of bones and bone tissue are determined. Subsequently these properties are correlated with the function of the structures or tissues. This study relates to the application of engineering concepts to biological phenomena. The understanding of how tissues and body structures behave is largely aided by using straightforward engineering test methods such as compression, tension, and torsion tests. The knowledge of the functional behavior of body tissues is a prerequisite for the development of implant devices and materials. In other words, the knowledge is also essential to improve upon those surgical procedures where temporary or permanent devices are used to aid or restore the function of the diseased or weakened skeletal part.

The above example from orthopedics based on experimental and clinical evidence can be easily generalized to one of the most fundamental aspects of biomaterials science: interactions between the implant materials and designs and the biological host do occur. As such, the interactions are an integral part of any biomaterials study, once the separate foreign and host systems are sufficiently characterized.

II. THE INTERDISCIPLINARITY OF BIOMATERIALS

In the second half of the 15th century it was possible for a man like Leonardo da Vinci to cover comprehensively all existing knowledge of science and philosophy. This remained so during the Renaissance and also later on until the end of the last century. It was not uncommon at all that creative and inventive individuals were active in different fields of endeavor.

Another example may be provided by the person of Thomas Young, who is remembered today mainly for his work on the wave theory of light and because the modulus of elasticity of materials is named after him. During his lifetime, he was not only a professor of physics, but also a practicing physician. In addition, he also wrote authoritatively on blood circulation, including wave propagation in arteries.

Today, the amount of knowledge available is so vast and the rate of increase so rapid that science has continuously split into smaller and smaller fragments; life sciences and engineering are today separately existing, well-developing disciplines of science which are characterized by largely differing subjects of study and very different problem solving techniques.

The separation caused by specialization has now assumed new importance. During the last 10 to 20 years, materials scientists and engineers have made considerable contributions to the understanding of physiological phenomena and the use of man-made materials for implants. There has been a strong stimulus for collaborative research between physicians or life scientists and engineers or materials scientists. However, at the same time the field became increasingly difficult for all those who had only limited training in the other disciplines involved.

This series of books entitled *Structure-Property Relationships in Biomaterials* attempts to alleviate this problem by dealing first with the basic aspects of materials science and engineering for a number of different materials: metals, ceramics, and polymers. Subsequently the fundamental properties of the biological materials are described *(Natural and Living Biomaterials)*. Only when this information is gathered is it possible to describe the direct interaction between implant materials and the host tissues. In *Functional Behavior of Orthopedic Biomaterials* behavior of orthopedic implant materials is thus covered. As such the aim of the present series is threefold:

1. To provide a basic insight into the factors which govern the functional behavior of the man-made materials used in surgery
2. To describe the properties of the host, important to the use of foreign materials in the human body
3. To assess and discuss the more important interactions of those implant systems where a sufficient degree of understanding has been acquired or where high-quality evaluation techniques have become standard practice

III. FUNCTIONAL BEHAVIOR OF BIOMATERIALS

What does constitute failure of any given foreign material in an implant application? Factors are many and can be traced back to a range of diverse phenomena. Without being comprehensive, possible mechanisms include wear or fatigue, failure of bonding, or untoward tissue reactions to the presence of the implant, corrosion, or static overload. Examples of these failure mechanisms are described herewith.

1. Wear: Teflon® acetabular components of total hip prostheses wear out and revision surgery to remove the prosthetic part is imposed.
2. Fatigue: Fatigue failures of bone fracture plates are frequently observed.
3. Bonding: present day fixation techniques rely on the use of poly(methylmethacrylate) cement; debonding either at the bone-cement or the cement-prosthesis interface is a precursory mechanism for implant loosening, eventually requiring revision as a result of clinical symptoms such as pain.
4. Tissue reactions: Polyethylene wear particles of the acetabular component and abraded bone cement particles are phagocytosed within the joint capsule; upon proliferation of this tissue into the cement bone interface loosening may occur.
5. Corrosion: The mechanisms of corrosion of implanted metals are fairly well recognized,

but the effect of corrosion products on local tissues or systemic functions is still a matter of debate, e.g., what causes the discoloration of the tissues in the neighborhood of titanium implants?

6. Overload: Bending of intramedullary nails shortly after implantation is indicative for overload failure. Overload failure can also occur without any deformation as is the case for fractured alumina components of total hip prostheses.

How can the performance and the success of materials implanted in the human body be analyzed? To answer that question, we consider what the finality of biomaterials science is. It is to apply materials in the clinical environment. In that respect, it has been suggested that the evaluation of the functionality is the more important criterion for assessing failure or success. In some clinical areas a high degree of clinical success has been reached. It is of a particular value to note that in those fields one is induced to wonder what the value of basic science might be. Does a basic science approach enhance the understanding of how successfully an implant functions? Apart from providing a sound basis for established procedures, it may help to improve existing materials and techniques. In addition, and equally important, the fundamental assessment of phenomena with one particular type of implant may enhance the understanding of the behavior of other devices. The example of acrylic bone cement in orthopedic surgery may clarify this point. The literature really abounds with studies on properties and structure-property relationships of acrylic bone cements. The data provided by these investigations have been very useful to determine the causes of loss of fixation with knee prostheses. Many of the early mechanisms of loosening knee prostheses were phenomenologically similar to loosening mechanisms of hip prostheses. While it was possible to observe characteristic loosening phenomena early on (about 2 years) with knee implants, it was only after 5 to 10 years that similar observations could be made with total hip prostheses. Thus, the fundamental understanding of poly(methyl methacrylate) (PMMA) properties led to the evaluation of mechanical failures of knee prostheses and eventually yielded prospective assessment of total hip prosthesis functionality.

The basic science approach can take numerous forms in biomaterials science. There are contributions both from the life sciences and from the engineering sciences. To name but a few, there are histology, histochemistry, histopathology, immunology, biochemistry, and electrophysiology for the life sciences; mechanics, ceramic engineering, polymer science, and metallurgy for the engineering sciences. One of the important questions in biomaterials science is who is going to provide the requirements these materials have to fulfill and who is going to relate fundamental life science aspects into materials design and engineering. The least one can say is that an understanding of each one's techniques, results, and limitations among the different disciplines is required. This does not necessarily mean that engineers should be able to perform, for example, histochemical or biochemical tests, but at least a basic level of understanding is necessary. However, one may be tempted to query this statement. Would it not be advantageous to combine the fundamental understanding of several sciences and preferably from life sciences and engineering within individual scientists? Would the insight into available knowledge and into the limitations of different disciplines not have a triggering effect on the advance of biomaterials science? Numerous examples of areas which could potentially benefit from this approach could be cited. For example, it is possible that there is an effect of proteins, lipoproteins, and enzymes on the ion release from metal implants. Experiments to address this hypothesis require an understanding of the physiological systems and of corrosion phenomena. The discovery of new concepts of controlled drug release systems requires knowledge of polymer and drug chemistry, the pharmacological action, and the nature of the diseases that may be treated. A third example is of total joint replacement where successful development of new designs depends upon a knowledge of patient factors as well as of engineering and materials science principles.

Major advances are still possible and only now do we begin to glimpse what may be ahead. This series of books intends to provide the basis for the interdisciplinary biomaterials research by reassembling the current understanding, both in the engineering and life science disciplines. First metals, ceramics, and polymers are discussed. In subsequent volumes, the tissues and the implant procedures are discussed with major attention on the interactions between materials and host and the processes which may be triggered by the insertion of a foreign material.

IV. STRUCTURE-PROPERTY RELATIONSHIPS OF FUNCTIONAL BIOMATERIALS

The failure mechanism as summarized above may be related to bulk aspects such as mechanical properties (fatigue, yield strength, ductility) or to surface-related properties (corrosion, biocompatibility, wear, etc.). The understanding of how synthetic materials behave in vivo and how they eventually may fail is intimately related, first to the development of new and improved materials and second to the design of better functioning devices and constructs. Analogous to materials science in its broadest sense, the basis is formed by microstructural theory. The understanding of all materials aspects, whether it is strength or fatigue failure resistance, wear tenacity, or corrosion resistance, is founded on a thorough knowledge of structure and properties on atomic size and microscopical level. Therefore this book starts off with a description of the organization of matter on both atomic size level and microscopic level. The relationship to macroscopic properties is outlined in different subsequent chapters dealing with each one of the separate material properties as they refer to implant surgery. Subjects which are thus treated are the strength and relationship to structure of metals and ceramics, the shape memory alloys, the principles of corrosion and electrode functioning, wear, surface coatings, and surface active materials.

Chapter 2

STRUCTURAL CHARACTERISTICS OF METALS AND CERAMICS

G. Heimke

TABLE OF CONTENTS

I. INTRODUCTION

The attempt to describe the basic properties of solids like metals and ceramics in close relation to their structures can start from different levels of information and knowledge about the essential features of the existence of matter. In order to achieve one unique common status, first, some simple and general information on atoms and ions and their interactions in compound formation is repeated. Solids are, then, introduced as "giant molecules". The different kinds of bonds existing in compound formation are followed further on into the field of solids where the basic properties distinguishing metals and ceramics are traced back to the kinds of binding processes prevailing in either of these two materials.

The concepts of solubility and corrosion are dealt with in some more detail because of their fundamental importance of material selection for implant application. Because of the steadily rising attention which electron exchange or redox processes is gaining for the interpretation and understanding of different kinds of biocompatibility, some basic concepts of the band theory of semiconductors are mentioned.

In discussing the main characteristics of metals, some emphasis was directed to stress the occurrence of a large variability of properties with materials of identical chemical composition. In addition, the plasticity of metals is explained in a little more detail. The main crystal structures occurring with metals and alloys are described as well as the main features of phases and phase diagrams of binary alloys. More detailed discussions of the alloys used most in medicine are left to the sections of this series particularly devoted to them.

The nonmetallic, inorganic materials (the ceramics) are introduced mainly from the viewpoint of the ionic structures from which most of the properties by which they are distinguished from the metals can be derived. As the wide range of the classical silicates containing ceramics cannot be considered for implant purposes because of their solubility and high impurity content, the discussion is concentrated on the materials based on pure metal oxides, the oxide ceramics, carbon-based materials, some nonoxide systems, and glasses or glass ceramics. The mechanical properties of these materials are evaluated in view of the discrepancy between the very high binding forces acting within the ionic lattices and the available mechanical properties. A much improved understanding of the details allowed for the development of methods of evaluation and testing with which the properties could be improved further.

In the concluding chapter some indications are summarized about the biological and clinical relevance of the basic structurally based characteristics of the materials concerned.

II. SOME PRELIMINARY REMARKS ABOUT SOLIDS

A. The Atom

All matter is made up of atoms. This old Greek concept was renewed during the last century and more and more convincingly confirmed during the last decades of that century and the first ones of the 20th century. During this century it was shown additionally that these atoms are by no means "indivisible", but that they have rather complicated structures, many details of which have become known step by step, a learning process which has, by no means, come to an end yet. However, sufficient features have already been established allowing the correlation of most of the technically relevant properties of matter back to the structural details of the atoms and, in this sense, to understand these properties.

Those details of the atoms which are needed to at least approximately understand the properties of matter, and in particular of metals and ceramics to be used in prosthetic devices, will be recalled briefly. The size of the atoms is of the order of 0.1 nm $= 1$ Å $= 10^{-8}$ cm $= 10^{-10}$ m, so a chain of 10^8 or 100 million atoms is about 1 cm long. Atoms have a mass of somewhat less than 10^{-23} up to some 100×10^{-23} g. They are composed of a nucleus of about 10^{-14} m = 10^{-5} nm diameter in which most of the mass is concentrated and which is surrounded by electrons which have a mass of only about 1/1836 of the smallest nucleus and a diameter of about 5.6×10^{-15} m = 5.6×10^{-6} nm. This nucleus is electrically charged (+), while the electrons carry an opposite charge (−). These electrical charges are multiples of the charge of the smallest nucleus occurring in an atom, the one of the hydrogen atom. The value of this unit charge is about e = 1.602×10^{-19} Å × sec.

In a first step of approximation, an atom can be described as consisting of a nucleus being surrounded by the electrons, the number of them being determined by the number of electrical unit charges carried in the nucleus. In this picture the negative electrons move around the positive nucleus on paths determined by the equilibrium of the attraction due to the electrostatic forces and the centrifugal forces. However, not all states of movement which are possible according to this equilibrium condition do occur in nature: only those are allowed *along which the length of one (circular) path of an electron around its nucleus times its* momentum (mass time velocity) is an integer multiple of Planck's constant.

In further pursuing this assumption and also considering the fact that the electron is rotating around itself (spinning), thus carrying an angular momentum itself, the arrangement of the electrons within particular "shells", the population of which corresponds to the number of elements in the rows of the periodic table, can be deduced.

If energies are exerted on atoms exceeding the binding energy of the electrons in the outermost shell, the electrons concerned can leave the atom. The remaining portion is lacking one negative electrical charge, thus carrying a surplus positive charge. Other atoms which are positioned to the end of a row in the periodic table can adsorb additional electrons, thus becoming negatively charged. Such atoms being either positively or negatively charged are called ions.

The existence of the rows of the periodic table, their lengths, and the details of the arrangement of the electrons in the different shells in the atoms can be understood using quantum and wave mechanics. Without going into the details of these theories, some further properties of atoms might be mentioned.

Atoms located in the first part of a row of the periodic table have their outermost electrons (valence electrons) bound relatively loosely, the energy for separating them from the rest of the atoms concerned being relatively small. So, these atoms are easily transformed into positive ions. This applies to, for example, H, Li, Na, Mg, and Ca. Those atoms, on the other hand, which are located more to the end of the rows, such as F, Cl, I, O, and S, tend to fill up their particular shells with additional electrons, thus becoming negative ions.

B. The Compounds

The formation of compounds of the type $Na^+ + Cl^- \rightarrow NaCl$ can, at least qualitatively, be easily understood from simple electrostatic considerations (heterologous bonding). However, there is still another process of the formation of compounds which can be understood if looked at from a different point of view: if it is assumed that the nucleus and the inside shells of an atom can be regarded as a unit, the "core", if looked at from the electrons of the outer most shell, and if it is further assumed that two such cores together can be jointly surrounded by the sum of their outermost electrons, this arrangement having a state of lower energy than must be ascribed to the two separate atoms, another method of the formation of compounds is established (homopolar bonding). The existence of molecules such as H_2, N_2, O_2, etc. can now also be understood qualitatively. In this picture the heterologous and the homologous bonding are the extremes of one phenomenon, the exchange of outer electrons between atoms.

However, there remains an important difference: the more the side of the ionic bonding is approached, the stronger the electrostatic contribution and the stronger and more rigid the electrons concerned are bound to their respective cores. This phenomenon will turn out to account for the major difference of the biological behaviors of most metals (freely moving electrons) as compared to many ceramics (consisting of ionic crystals).

There is still another point of view regarding the bonding of two or more atoms: if the electrons which are shared between the atoms concerned are more or less strongly localized directionally, e.g., in one particular position (orientation) between the atoms, another type of bonding results (covalent bond). The carbon atom is particularly susceptible for this type of bonding; the most pronounced example is the diamond.

C. The Solids

With the concepts of atoms, ions, homologous and heterologous bonds, and a basic understanding of the existence of compounds, a first approximation of a description of solids can be approached. For this purpose the solids are subdivided into four groups:

1. Mainly heterologously bound solids (e.g., ionic crystals)
2. Mainly homologous bound solids (metals)
3. Mainly amorphous materials (glasses)
4. Plastics
5. Composites (bone, steel concrete)

Most solids, in particular those of the first two groups, can be regarded as agglomerates of atoms or ions like molecules, but consisting of an enormously larger number of them, so being giant molecules, containing about $10^8 \times 10^8 \times 10^8 = 10^{24}$ atoms per cubic centimeter. So, the basic concepts used in understanding the formation of molecules and compounds can be used in trying to understand the formation and basic properties of solids.

There is one more aspect which all solids have in common and which also can already be found in molecules. This is the existence of ordered structures, particularly directional arrangements of the atoms, ions, or molecules of which the solids are composed. This ordering phenomenon can qualitatively be understood remembering the following experiment: a rectangular glass vessel may contain hard balls of equal size. The number of balls shall be chosen so that they are to form about five or six layers if at rest and occupy not more than half the volume of the vessel. If the vessel is vibrated with a suitable frequency and high enough amplitude, the balls will fill the available volume nearly homogeneously, so simulating the high temperature or gaseous state. If, with a flash camera, a photograph is taken, no ordering will be detectable (Figure 1). In decreasing the amplitude of the vibration, the density of the balls in the lower portion of the vessel will increase and a state can be reached in which a photograph, taken as mentioned before, will show more and more groupings of the balls in arrangements like triangles and hexagons, here and there, but disordered areas in-between (Figure 2). This kind of arrangement is called "short-range ordering" and is characteristic for most liquids and amorphous solids. If the amplitude of the vibration is made still smaller (simulating a further decrease in temperature), the degree of ordering is increasing until the balls are closely packed in layers, only the sequence of the arrangement of one layer upon the other may vary (stacking fault), and there will be some other irregularity here or there (lattice defects) as can be seen in Figure 3. So, this simple model in which the atoms are simulated by spheres and, therefore, geometrical influences of one particular kind are considered only and can account for quite some ordering phenomena in particular in metals.

However, there are other contributions to ordering. If the material concerned is not composed of electrically neutral atoms or molecules, but of electrically charged entities like ions, the requirement of electric neutrality within short distances gives a strong contribution to ordering. This contribution can be so strong that structures which would not be stable from the purely geometrical point of view are realized as in some salts or in most ceramics.

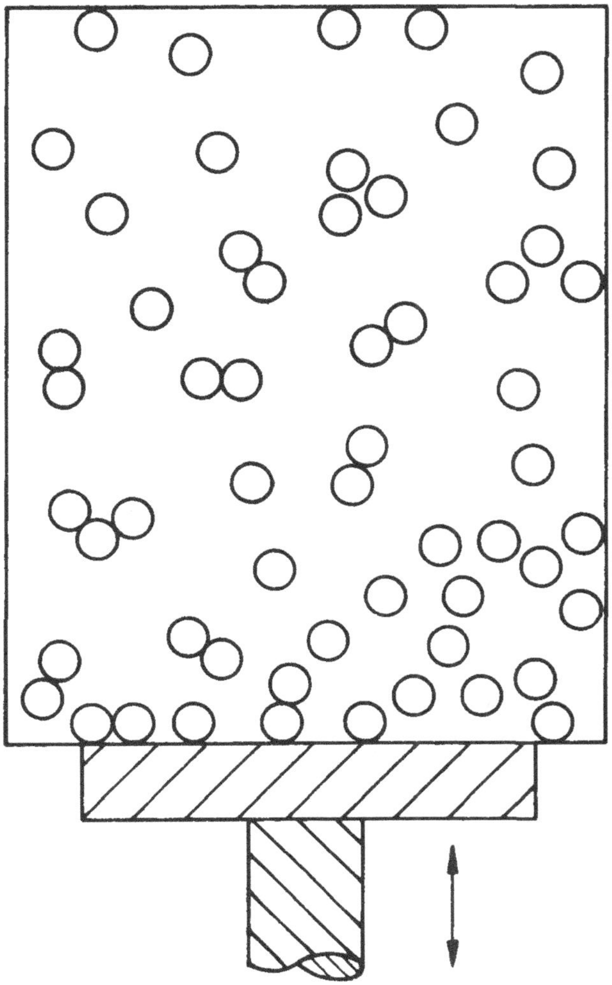

FIGURE 1. Schematic representation of an experiment simulating the gaseous state: steel balls in a vibrating glass vessel. Frequency and amplitude (energy) chosen so that the balls fill the volume nearly homogeneously.

A further contribution to ordering is a particular property of the arrangement of the electrons in the atoms concerned like in carbon and silicon. In diamond (see also Section IV.C), each carbon atom is symmetrically surrounded by four nearest neighbors exchanging one of its four valence electrons with each one of them. Within these C=C double bonds, the valence electrons are strongly bound and localized. Here, again, arrangements of the atoms in space can result which would not be stable from purely geometrical considerations. The example of the diamond even shows that the strong binding forces can result in extremely stable structures as expressed by its extreme hardness.

All such structures with a degree of ordered arrangements exceeding the nearest neighbors are called crystals, more specifically they are called single crystals. The vast majority of solids is composed of crystals, but usually a given solid is not one single crystal; it is rather made up of a larger number of closely packed crystals of different sizes (polycrystalline materials).

Solids, which are not composed of crystals, are called amorphous. The glasses belong to

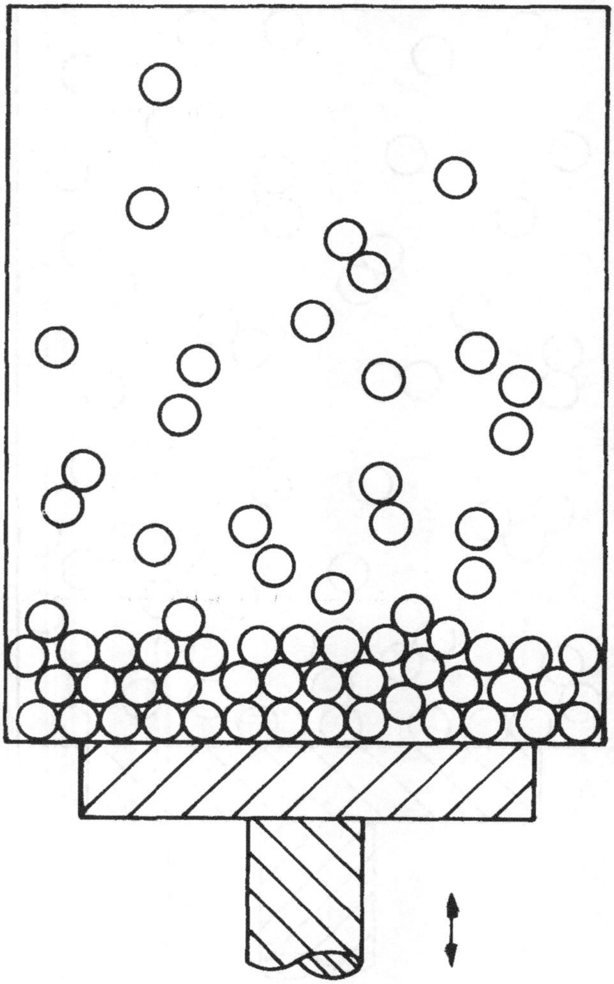

FIGURE 2. Schematic representation as in Figure 1, but with reduced amplitude of vibration (reduced energy): formation of short-range ordering.

this group. However, in them there is still some short-range order, as in most liquids. Therefore, glasses can be considered as solidified liquids.

An essential feature of all these long-range ordering phenomena is their repetition in all three dimensions; all these ordered structures can be deduced to one single entity, called unit cell, of which every crystal can be built by repeating the structural arrangement of the unit cell in all three directions in space. The resulting three dimensional structure is called "lattice" or crystal lattice. Figure 4 shows the most simple types of unit cells and mentions examples of materials in which they are realized.

In order to understand many basic properties of solids one more concept is necessary. The distinguished points of the lattice define the positions of the atoms, ions, or molecules of which the materials are composed if and only if the crystal is in complete equilibrium. However, this state can not be reached under normal conditions: thermal energy excites the lattice elements to oscillate around their equilibrium positions with a statistical distribution of the vibrational energies. If the forces acting on an atom during these thermally stimulated oscillations exceed the forces with which it is held in or bound to its lattice position, the

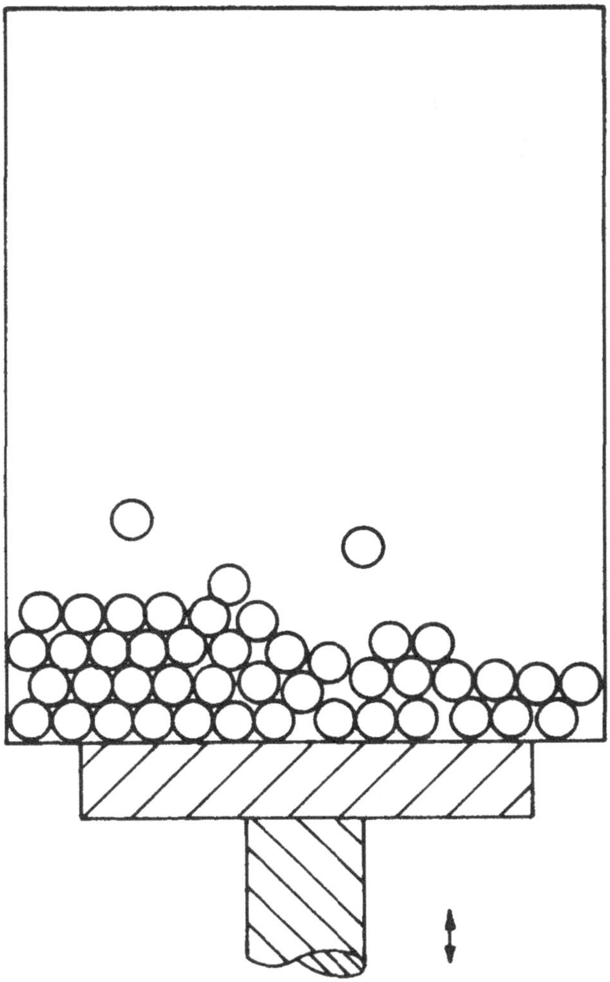

FIGURE 3. Schematic representation as in Figure 2, but with a further reduced amplitude of vibration: most of the balls can be found in an ordered arrangement, but there are some ''defects'' in this ''lattice''.

lattice element concerned can leave its equilibrium position. Its original lattice position thus becomes empty, a ''vacancy'' is created, the atom concerned is either sitting in between other lattice points as an ''interstitial'' or it finds another empty place into which it can jump (Figures 5 and 6). So any crystal does contain a basically temperature dependent vacancy concentration allowing for ''jumps'' and, thereby, for diffusion.

Such a single isolated vacancy belongs to the group of ''point defects''. Other point defects are foreign atoms or ions which have entered the crystal lattice ''erroneously'' as contaminants. If they have been built into the lattice deliberately, they can be regarded as having been soluted into the material, so a ''solid solution'' has been formed.

Under particular conditions point defects can agglomerate into ''clusters''. If the point defects consist of foreign atoms or ions and if the clusters become more and more densely populated, in many cases these atoms or ions start to form their own lattice if this is more favorable energetically. So, a ''precipitate'' of a foreign lattice is built inside a host crystal. If the point defects are vacancies, the clusters mostly arrange themselves as chains, thus (Figure 7) interrupting a ''lattice plane'' creating an ''edge dislocation''. It is this type of

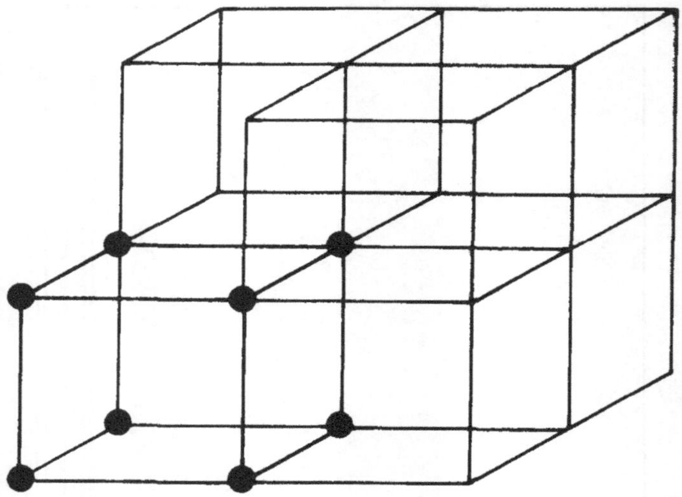

FIGURE 4. Simple or primitive cubic lattice with one unit marked (''unit cell''). A unit cell is usually shown with all the atoms or ions, the centers of which are located inside the cell. However, the number of atoms or ions per unit cell is much lower because each corner of a cubic lattice belongs to eight cubes; one eighth of all corners can be attributed to each cube only. Thus, in a simple cubic lattice, there is only one atom or ion per unit cell. An atom or ion located on an edge between the unit cells belongs to four cells simultaneously in the cubic lattice, thus, only a fourth of them can be attributed to one cell. A similar argument holds for atoms or ions located in planes separating two unit cells so that only half of them can be attributed to one cell. This simple cubic lattice is realized in NaCl with centers of the Na^+ and Cl^- ions at the corners alternatively in all three directions.

agglomeration of vacancies which allows for the plasticity of the solids, particularly metals. Figure 8 shows schematically how a plastic deformation can be explained by the movement of a dislocation.

D. The Different States of Matter, Phases, and Phase Diagrams

In the description of the experiment (see Section II.C) in which a glass vessel containing balls had been vibrated, situations had been mentioned simulating the gaseous state (complete disorder), the liquid state (higher density and partial order), and an ordered state (solid). The change from one of these states into the other was caused by changing the energy of vibration supplied to this vessel, so simulating a temperature change. Such different states (kinds of arrangement) of the same matter (the balls in our experiment) are called ''phases''; the change from one state into another is called ''phase transition''.

The most common phase transitions are the liquification of a gas with falling temperature or the evaporation of a liquid with rising temperature characterized by the boiling temperature and, analogously, the solidification of a liquid (freezing), or the melting of a solid at the melting temperature (both these characteristic temperatures being pressure dependent). Table 1 gives a list of the boiling and melting temperatures of the most commonly used materials of interest for endoprostheses.

If, after solidification, the temperature is decreased further and further, one or more changes of the crystal structure can be observed. These do also occur at particular well-characterized temperatures. They are also called phase transitions. So, the term ''phase'' is also used for the different crystalline states which a particular material can adopt at different temperatures.

The occurrence of different crystalline phases within different ranges of temperature can be understood qualitatively by somewhat extending the geometric model mentioned above (Section II.C). If two kinds of atoms of different weight and size are considered, the vibrational (thermal) energy will result in different amplitudes of their oscillations. The mean

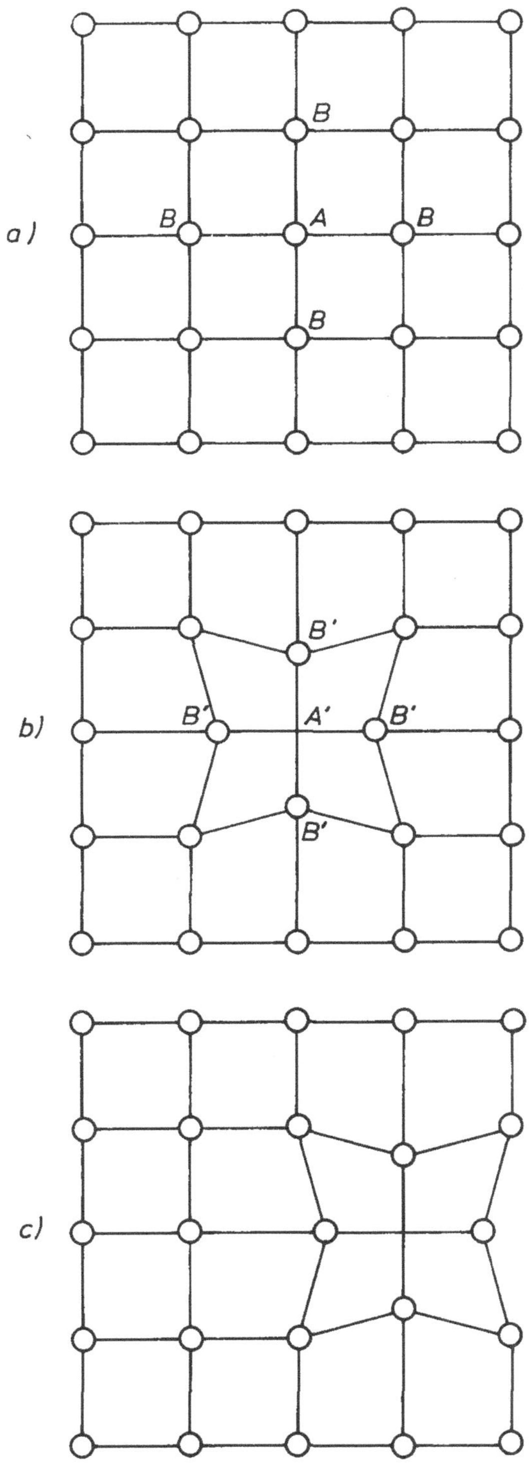

FIGURE 5. An atom, A, removed from its lattice position in (a) creates a vacancy A′ and a distortion of the neighboring atoms B (b). If one of the B ions is jumping into the empty space, a new vacancy (and lattice distortion) is created (c). Thus, vacancies can move through the lattice.

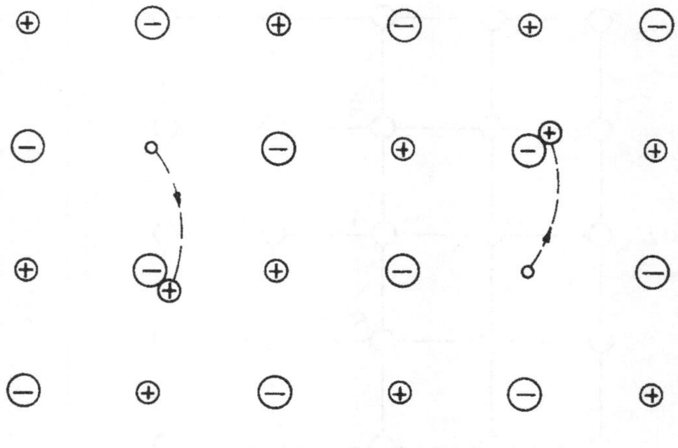

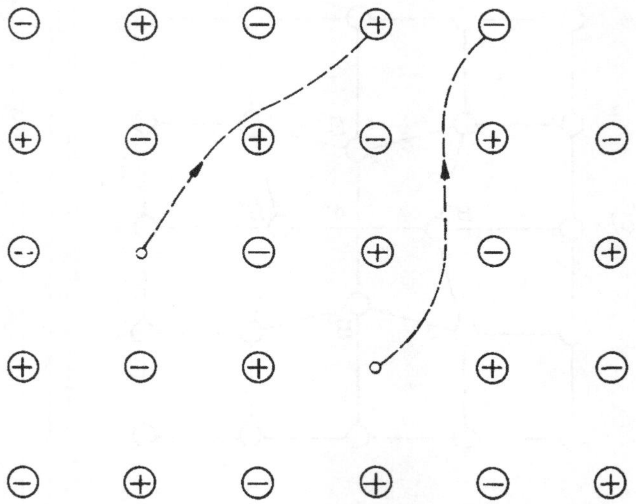

FIGURE 6. An atom or ion leaving its position can create a vacancy and can move to a site usually not occupied in the lattice concerned. There, it is called an interstital (A). Vacancies can also be created if atoms or ions can move to the surface or a grain boundary (B). In ionic crystals, such vacancies will usually be electrically charged to maintain electrical neutrality (C).

value in time of the displacement from their equilibrium position will, of course, also be different. However, if the temperature (energy of vibration) is changed, the amplitudes of the oscillations of the two kinds of atoms will change differently and with it the time mean value of their displacement. This mean value, however, determines the space the atom concerned must have available within the lattice. So, with a changing ratio of the mean values of the two kinds of atoms, another geometric arrangement may become more favorable energetically and, therefore, result in a solid-state phase transition.

It is generally known that the phase transitions from the gaseous state to the liquid and from there to the solid can be "delayed", giving rise to supercooled gases or liquids. The opposite also can occur, resulting in superheated solids or liquids. The situation in which a

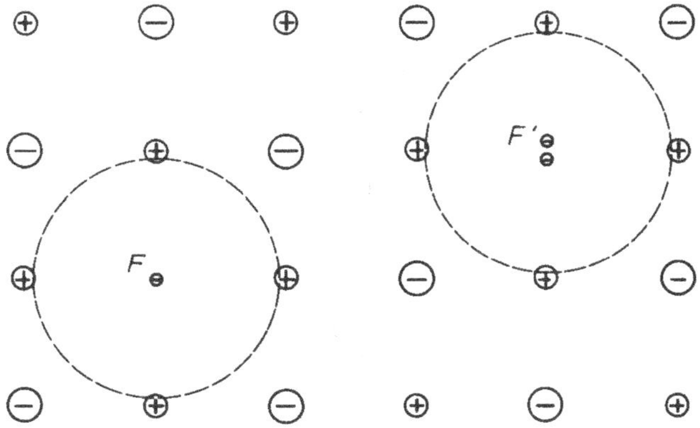

FIGURE 6C.

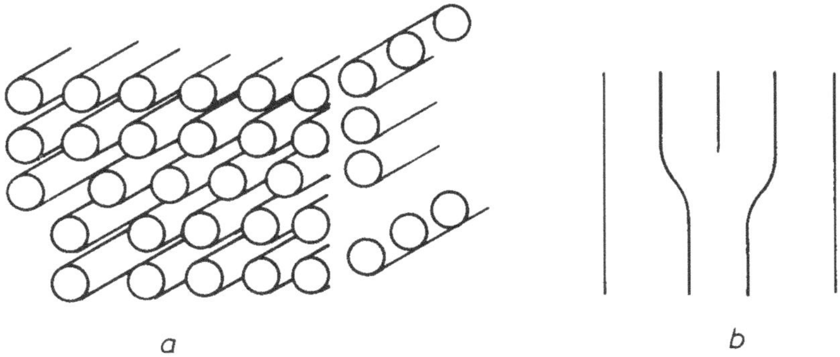

FIGURE 7. Schematic representation of an "edge dislocation" (a) and a still more abstract version showing the lattice planes as lines only (b).

material is outside its equilibrium state is called "metastable". The phenomena of supercooling and superheating and the existence of metastable phases is also known for phase transitions of solids.

The transformations from the gas to the liquid and from there to the solid state involve energies. The same is true for the solid-state phase transitions in which a rearrangement of the crystal lattice is involved. The model with the vibrating balls mentioned in Section II.C is completely inadequate because it does not allow for the occurrence of these "heats of transition".

E. Basic Properties of Solids

The properties of solids can be related to all the constituents by which a solid is defined with different degrees of influence on any of the properties under consideration.

1. Of course, the kind of atom or atoms contained in a given solid is of basic importance, as well as the state in which they are entering, for example, atoms or ions of different degree of ionization.
2. The kind of bonding can be different as in diamond and graphite or in quartz and silicate.
3. With the same atoms and basically the same type of bonding, their arrangement in space can be different, so different crystal phases will result in different properties.

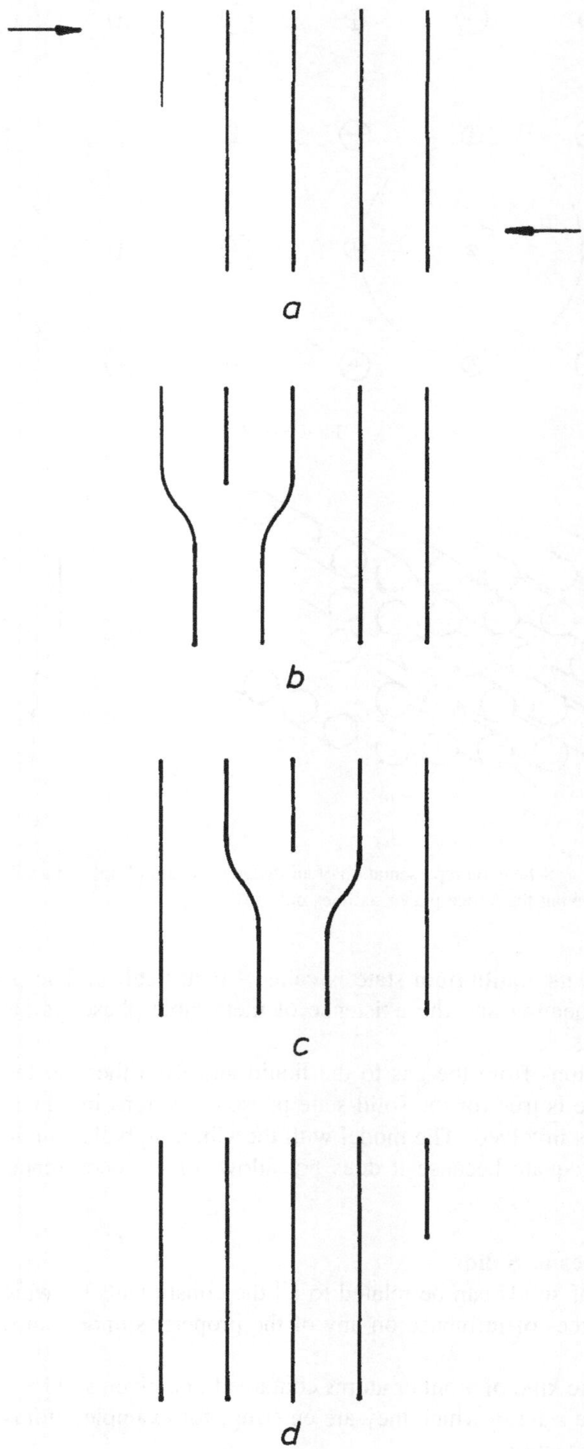

FIGURE 8. Using the most schematic representation of Figure 7b, the movement of a dislocation and the resulting deformation is demonstrated going from portion (a) to (d).

Table 1
BOILING AND MELTING TEMPERATURES OF MATERIALS USED IN ENDOPROSTHESES

Material	Boiling temp. (°C)	Melting temp. (°C)	Ref.
Stainless steel	—	1520	45
Co-Cr-Mo alloy	—	1350	45
Titanium	3220	1680	46
Ti-6Al-4V	3220	1650	46
Ti-5Al-2.5 Fe	3220	1650	46
Al_2O_3 (corundum)	2980	2050	
Bioglass	—	1450—1475	
Carbon	4827[a]	3550	

[a] Sublimes above 3500°C.

4. Many solids are not only composed of one single type of crystal, but of several. Their relative ratios also influence the properties of the solid concerned.
5. Two single crystals being chemically identical and having the same crystalline phase can still differ in the type and density of the lattice defects they contain.
6. The dimensions of the crystals within a solid can vary considerably; the resulting "grain size distribution" can be of major importance for several properties of a given solid.
7. The shape of the crystals can be varying or homogeneous; their relative orientation can show some directional preference ("shape texture").
8. If the lattices of the crystals within a solid do show some preferred orientation, deviating from the random directional distribution, the solid is said to have a "crystal texture".

For many purposes it is convenient to distinguish between those properties which are mainly determined by the crystalline phase (or phases) of the solid concerned and which are, if at all, only slightly affected by lattice defects, grain size variations, or shape texture as opposed to the group of properties strongly dependent on lattice defects, contaminations of any kind, and/or crystal texture. The first group (called structural-dependent) comprises the specific heat, elasticity (Young's modulus), thermal expansion, compressibility, heat of formation, some main optical properties, dia- and paramagnetism, and the electrical conductivity in metals.

The properties mainly determined by the defect structure are all diffusion-related effects, ion conductivity and the electrical conductivity of all semiconducting materials, and most aspects of mechanical strengths.

In applying solids generally, not all the properties of a given material are made use of, but there is rather a limited number of properties which are of interest for a particular purpose. So, it is mostly sufficient to consider the combination of those interesting properties, disregarding the rest. Thus, the magnetic and electrical properties need not be considered for most bone and joint replacements, neither the heat conductivity and the color of the materials concerned. The materials used for kerato prostheses or middle ear implants are not exposed to higher mechanical stresses and mechanical fatigue is not an important issue relating to them.

However, for all materials which are intended to or might eventually come in contact with parts of the patients' bodies, their compatibility with body tissues and fluids is a prerequisite. A basic requirement, therefore, is the nonsolubility because, of course, any product of solution (solute) must be expected to influence the cell activity of the host body

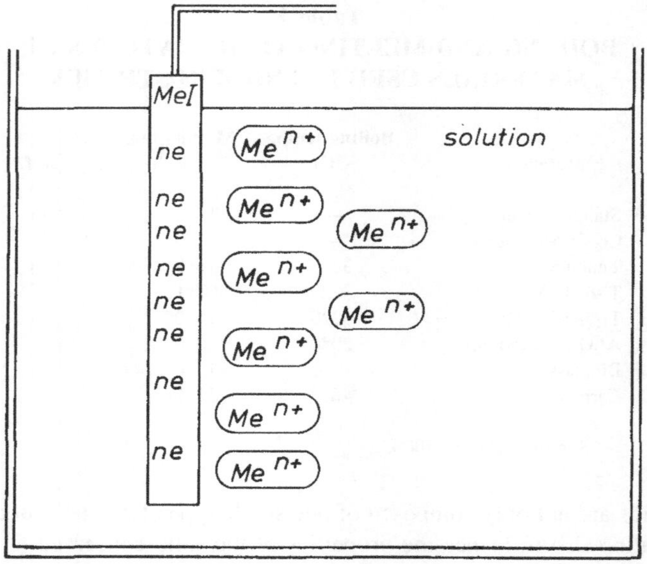

FIGURE 9. Half cell of a metal (Me) dipped into a solution with metal ions (Me^{n+}) soluted and the equivalent number of electrons (ne) remaining in the metal. The metal (Me) is, thus, negatively charged as compared to the surrounding solution.

and these influences are, in the vast majority of cases severely detrimental. In addition, mechanical strength is a basic requirement for nearly all the implant materials. This is obvious in most orthopedic applications, remembering the rule to sacrifice as little healthy tissue as possible for the anchorage of, for example, the components of a joint prosthesis. In other cases weight is a critical factor, as in heart valves or pace makers, limiting cross sections or wall thicknesses.

F. Solubility and Corrosion

By definition, any phase containing more than one component is a solution. The component present in excess is usually called the solvent, the other or the others being called solute or solutes.

For the purpose of the consideration of the biocompatibility and corrosion of implant materials, the discussion can be confined to the immersion of solids into an aqueous solvent. As far as metals and ceramics are concerned, the solutes will be ions, so this is a problem of electrochemistry.

From the point of view of electrochemistry, a solid dipped into a solvent is said to constitute a ''half cell''. If two suitable half cells are connected, one obtains an electrochemical cell.

One of the simplest types of half cells is realized if a metal electrode is in contact with a solution containing ions of the metal. It can be represented as $Me/Me^{n+}(c)$, where c is the metal-ion concentration, n is the number of unit charges carried by this metal ion, and the vertical bar denotes the phase boundary between the solid and the liquid concerned. The reaction occurring at this boundary is the solution or deposition of the metal, according to $Me = Me^{n+} + ne$, where e is the negatively charged electron.

From an atomistic point of view, this process can be described using Figure 9 in which a metal dipped into a solution is shown. From the surface of the metal Me ions, Me^{n+} are released into the solution in a similar manner as atoms are evaporated from the surface of a liquid into a gas or are sublimated from a solid into a surrounding gas. However, what

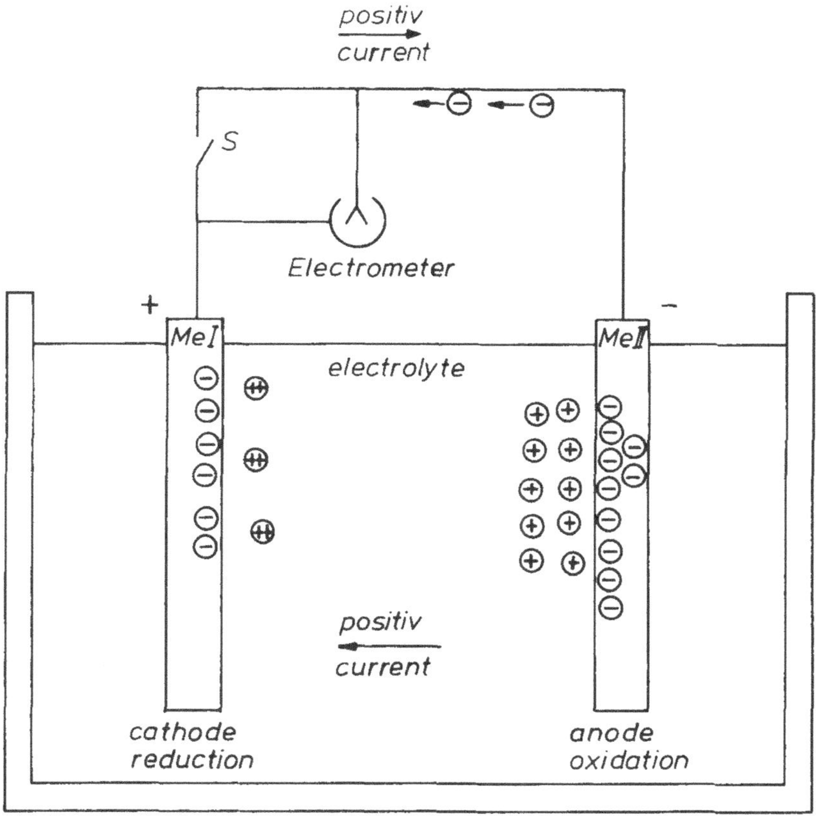

FIGURE 10. Electrolytic cell (schematic) with metal electrodes, Me I and Me II, an electrolyte, outside wires connecting an electrostatic electrometer, and a switch S.

goes into solution are not atoms, but electrically charged ions, their electrons remaining with the metal. So, an electrical field is established creating a force pulling the metal ions back towards the surface of the metal. Thus, in equilibrium, a "cloud" of Me^{n+} ions will be present in the solution along the surface of the metal.

If, for example, the solution contains a solute R which can react with the Me^{n+} ions resulting in a highly insoluble compound MeR according to the net reaction $Me^{n+} + R \rightarrow MeR - ne$, for which the electrons are supplied from those within the metal, then, the equilibrium will not be reached as long as the solute R is available. Thus, the metal will go into solution continuously and will be corroded. This is the basic process of metal corrosion.

If a second metal is dipped into the same solution some distance apart from the first one and if no such solute R is present, both metals (electrodes) can reach their equilibrium separately as long as they are not connected electrically. Figure 10 shows this situation schematically, including an electrostatic potentiometer indicating the existence of an electrical potential between the two electrodes as long as the switch S is kept open. The one consisting of metal II containing more electrons than the other of metal I is negatively charged as compared to Me I. In such a situation it is said that the metal II is anodic as compared to metal I which is called cathodic. The potential is said to be positive (expressed in volts) for metal II in comparison to metal I. This direction of the definition of positive and negative can be understood if looked at from inside the solution: the cloud of ions in front of the electrode of Me II is more positive (anodic) than the one in front of Me I which, in turn, appears to be less positive (cathode). If the switch S in Figure 10 is closed, the (positive) ions will move from metal II to metal I inside the solution in accordance with the general definition of a positive current. In the outside wires the electrons will move from electrode II to electrode I, also constituting a positive current, but in the opposite direction so that the current within the complete loop (consisting of the series of outside wires, electrodes, and solution) will remain zero as required by basic laws.

In the arrangement shown in Figure 10, there will be a net flow of metal II to the side of metal I if the switch S is closed. The Me II ions arriving at the side of metal I will be discharged there, the electrons needed for this reduction arriving via the outside wires. Thus, the metal II will be dissolved; it is electrolytically corroded.

Pairs of all solids can be arranged in a manner as described above, and the electrical potential between them can be measured. For gases, a porous tube of an extremely inert metal like platinum can be used through which a flow of the gas is passed, the potential being measured against this tube. If one material is arbitrarily defined as having a particular potential, all others can be characterized by their voltage against this one reference material. The material usually chosen is hydrogen H^+ to which a zero potential is ascribed. Table 2 gives an example of such electrode potentials.

From the list in Table 2, it can be deduced that, for example, titanium has a potential of +4.4 V against silver using the following calculation:

$$\begin{array}{rcll} Ti & = & Ti^{3+} \; + 3e \rightarrow & \qquad 2.0\ V \\ \underline{(-)\ 3Ag} & \underline{=} & \underline{3Ag^{+} + 3e \rightarrow 3 \times 0.8 =} & \underline{-2.4\ V} \\ Ti - 3Ag & = & Ti^{3+} \; - 3Ag & \qquad 4.4\ V \\ Ti + 3Ag & = & Ti^{3+} \; + 3Ag & \end{array}$$

The positive potential corresponds to a negative free energy ΔF, indicating a spontaneous reaction to the right side. Thus, from this point of view, titanium must be expected to be corroded under these conditions. However, this is not observed. The reason for this different behavior can be found in the formation of a dense and stable oxide layer on the surface of this metal, which is said to be passivated by this kind of oxide formation.

The basic process of passivation is the formation of surface conditions slowing down any electrode reaction. While the potentials discussed above can be regarded as supplying the driving forces of corrosion, the concept of passivation is related to the kinetics involved.

The potentials can be measured only in an equilibrium situation, which is usually realized with open circuit conditions. If, in a cell like the one shown in Figure 10, the switch S is closed, the ions of Me II flow to the electrode of Me I, the Me I ions in front of which are "driven back". At the same time, however, all other ions in the solution concerned will start moving, like the H^+- and O^{2-} ions. So, an additional deviation from the equilibrium condition is caused. The oxygen ions arriving at the anode can either be neutralized and remain in solution

$$2O^{2-} \rightarrow 2O + 4e$$

or bubble out of the solution.

$$4OH^- \rightarrow 2H_2O + O_2\uparrow + 4e$$

Writing the reaction in this way shows that the solution in front of the anode might change its degree of acidity towards a lower pH value. This can also occur via the reaction

$$Me + 4\ OH^- \rightarrow MeO_2 + H_2O + 2H^+ + 6e$$

Table 2
STANDARD OXIDATION POTENTIALS (IN VOLTS) OF SOME MATERIALS (WITH H^+ = 0)

Element			Potential
Li	=	$Li^+ + e$	3.045
K	=	$K^+ + e$	2.92
Ca	=	$Ca^{2+} + 2e$	2.87
Na	=	$Na^+ + e$	2.71
Mg	=	$Mg^{2+} + 2e$	2.37
H^-	=	$H + e$	2.25
Ti	=	$Ti^{3+} + 3e$	2.00
Al	=	$Al^{3+} + 3e$	1.66
Ti	=	$Ti^{2+} + 2e$	1.63
Zr	=	$Zr^{4+} + 4e$	1.53
Mn	=	$Mn^{2+} + 2e$	1.18
V	=	$V^{2+} + 2e$	1.18
Zn	=	$Zn^{2+} + 2e$	0.763
Cr	=	$Cr^{3+} + 3e$	0.74
Fe	=	$F^{2+} + 2e$	0.44
Co	=	$Co^{2+} + 2e$	0.277
Mo	=	$Mo^{3+} + 3e$	0.2
$1/2\ H_2$	=	$H^+ + e$	0
Cu	=	$Cu^{2+} + 2e$	−0.337
Ag	=	$Ag^+ + e$	−0.799
Pt	=	$Pt^{2+} + 2e$	−1.2
Cl_2	=	$2\ Cl^- + 2e$	−1.36
Au	=	$Au^{3+} + 3e$	−1.50

resulting in the formation of a metal oxide, thus changing the situation of this cell severely. If a dense oxide film is formed on the surface of the anode metal and if this oxide is insoluble under the conditions concerned, the current density will drop considerably approaching values which can be considered negligible.

If, by using an outside power supply, the anode is made still further charged, this oxidation process can be carried on creating still thicker and more stable oxide layers if the conditions are suitably chosen. This process is used technically to get stable (and even colorful) surface oxides on aluminum. There seems to be such an effect in vivo with titanium implants without any foreign voltage or power supply.[1,9] If, however, the oxide formed on the anode is not depositing on the metal surface, but is removed continuously (by whatever process), a severe anodic corrosion is taking place.

Along the cathode, electrons are consumed, so reactions such as:

$$Me^{n+} + ne \rightarrow Me$$

or

$$4H^+ + 4e \rightarrow 2H_2 \uparrow$$
$$Me + 2H_2O + e \rightarrow MeOH + OH^- + H_2 \uparrow$$

can occur, increasing the pH of the solution.

Which of the two processes, a more anodic or the more cathodic controlled one, will actually occur in a particular case can hardly be predicted theoretically. Experimentally, curves of the current density vs. the potential can be measured and plotted as shown schematically in Figure 11. The type of prevailing corrosion process can, then, be determined, using the Evans diagrams as explained in Figure 12. The passivation phenomenon can also be detected from such diagrams as outlined in Figure 13.

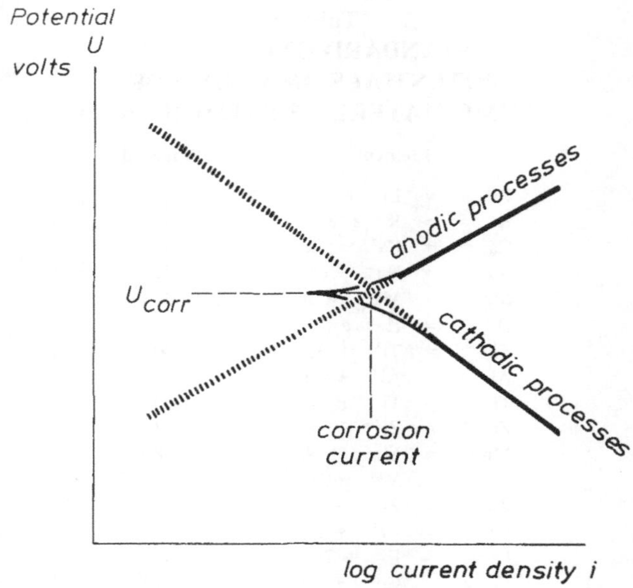

FIGURE 11. Schematic of potential, U, vs. log current density, i, curve, polarization curves, produced by making the metal surface measured an anode, and, then a cathode. The extrapolations of the anodic and cathodic curves (dotted portions) produce the diagrams (Evans) of Figure 12. (From Krüger, J., *Corrosion and Degradation of Implant Materials*, American Society of Testing Materials, Philadelphia, 1979, 107. With permission.)

If the potential as measured against a far away electrode (isolated electrode potentials) is plotted against the time, curves like those shown in Figure 14 can be obtained also allowing a judgment on the corrosion behavior. However, it must always be kept in mind that all such results are only valid for the experimental conditions chosen and that it is very difficult to reliably simulate the actual in vivo situations.

Solubility and corrosion are by no means limited to metals. As far as solubility and corrosion are electrochemically controlled, they are however, mainly confined to materials showing at least some electrical conductivity. As most of the plastics and oxide ceramics used in surgery are very poor conductors or rather good or extremely good insulators, their corrosion is more solution, reaction, and/or diffusion controlled.

Most of the plastics are polymers, the large molecules of which are interconnected or bridged by additives. In most cases it is the leaking of these additives into the surrounding tissue that causes disturbances within the body and/or leads to a change of the properties of the solid concerned. The rate controlling process can, of course, be either the bulk diffusion coefficient, the solubility or grain boundary, and/or surface diffusion properties. Reactions and their kinetics between the different constituents of the solid concerned and body molecules must also be considered. However, there are not yet any general rules available for either measuring or predicting the corrosion of these materials in body environment.

For the main representative of oxide ceramics used in surgery, the Al_2O_3 ceramic, corrosion phenomena are confined to grain boundary diffusion with ceramics of lower degrees of purity.[2,41] Ca ion concentration phenomena have also been reported for ceramics containing Ca impurities.[36,37] The glasses and glass ceramics which have been found to be compatible with bony tissue and even stimulate this tissue to form a particular kind of bond owe this property to a solution process during which calcium and phosphate ions are leached out.[15,20] Thus, in this case, corrosion is a positive property at least as long as it does not continue beyond the formation of the desired bond.

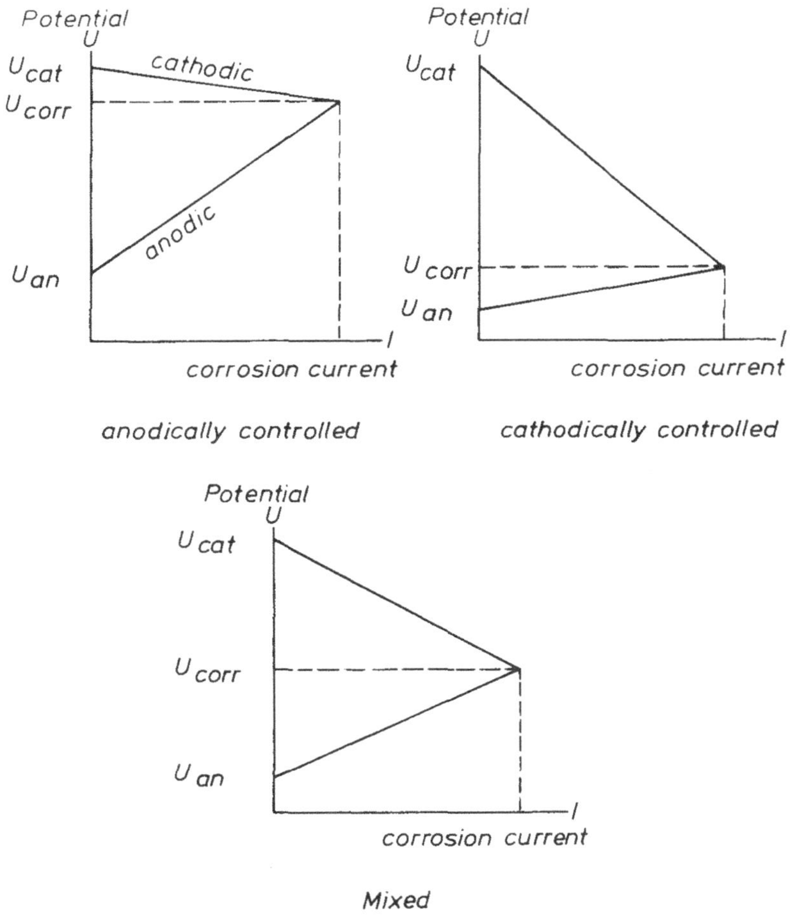

FIGURE 12. Diagrams (Evans) showing three ways in which the corrosion processes can be mainly determined. U_{an} and U_{cat} are, respectively, the equilibrium (reversible), anodic, and cathodic potentials. (From Krüger, J., *Corrosion and Degradation of Implant Materials*, American Society of Testing Materials, Philadelphia, 1979, 107. With permission.)

Another group of materials for which the dissolution is a favorable property is the resorbable ceramics. These calcium phosphates and hydroxyapatites have been shown to allow for a particularly close bone formation along their surfaces under suitably chosen conditions.[12,31] The rate of their dissolution was found to be controlled by several production parameters,[18] most of which can be understood to influence the stability of the lattices concerned, but the details of the dissolution processes are not yet generally agreed upon. Some experimental results indicate a solution process which destroys the necks between the particles, the thus freed particles can be annihilated by biological processes (phagocytized). As far as the more dense materials of this type are concerned, recent results indicate an ion exchange process by which the Ca^{2+} ions are replaced by Na^{+} ions, resulting in severe material destruction.[38]

Of course, all the dissolution and/or corrosion processes mentioned above do contribute to all the aspects of material fatigue, but the reasons for fatigue are by no means confined to corrosion. There are many more processes contributing to the fatigue phenomena, in particular the dislocation movements.

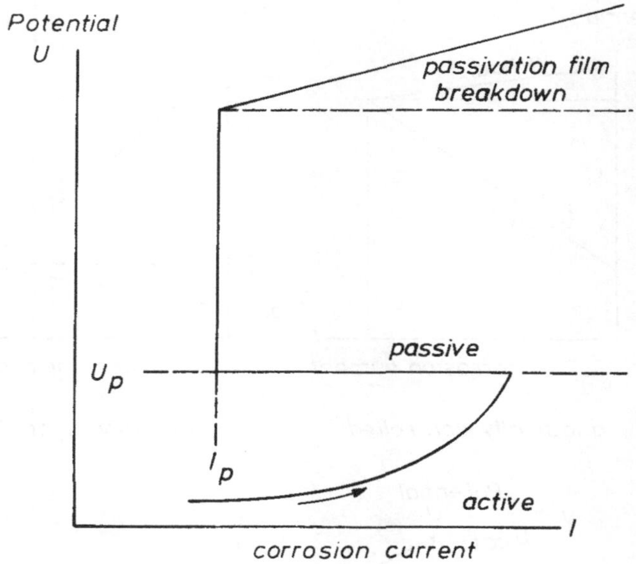

FIGURE 13. Anodic polarization curve (schematic) with three different potential regions. At the potential, U_p, the system becomes passive with the passive current density, i_p. (From Krüger, J., *Corrosion and Degradation of Implant Materials*, Philadelphia, 1979, 107. With permission.)

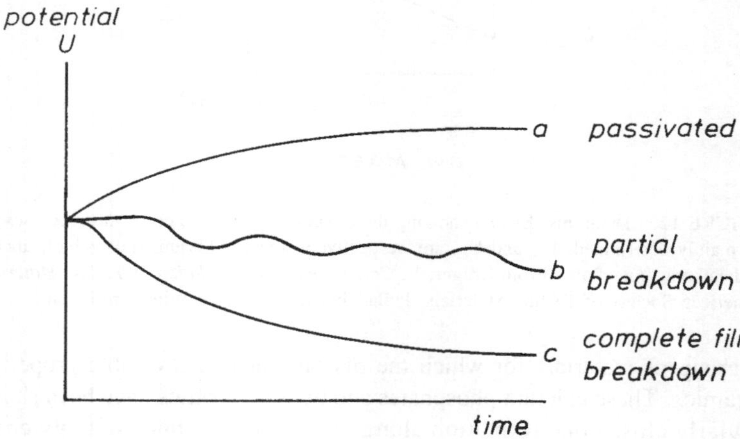

FIGURE 14. Time dependence of the isolated electrode potentials for a metal with an intact film (passivated), a metal exhibiting film breakdown after initial thickening, and a metal that shows film breakdown after immersion. (From Williams, D. F., *Am. Rev. Mater. Sci.*, 6, 237, 1976. With permission.)

G. Electronic Energy Distribution in Solids

Some observations regarding the blood compatibility of implant materials have led to studying the influences of the conductivity of some semiconducting materials, mainly oxides, on blood clotting.[34] This aspect might be found to contribute to the obvious tissue compatibility of some oxides or oxide films.[13] Therefore, some basic concepts used to describe the properties of semiconducting materials will be explained qualitatively.

Macroscopically, all materials having an electrical conductivity about halfway between

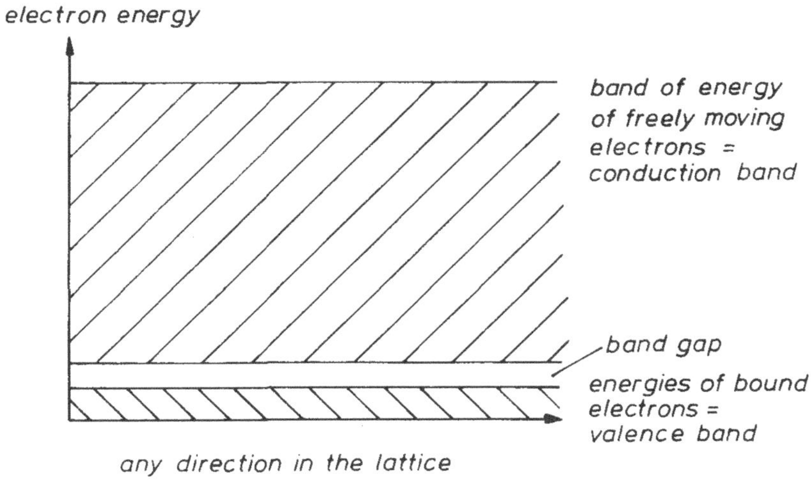

FIGURE 15. Band representation of energy states in solids.

the (highly conductive) metals and the extremely well-insulating materials like most ceramics are considered to be semiconductors. For these three types of materials a theoretical understanding of their electrical behavior can be given which clearly justifies this grouping into conductors, semiconductors, and insulators. Of course, there is a continuous chain of materials in between these groups, but the theoretical model will allow for an explanation of all these intermediate cases, too.

The model uses two basic ideas: first the energy of those electrons which can move rather freely within the crystal lattice and might even be able to leave it, if other electrons are entering the lattice at other places to maintain the electrical neutrality of the whole crystal or solid concerned, and, second, the very basic principle (of Pauli) saying that, within a given system, no energy state will be occupied by more than one electron. For the freely moving electrons, the whole single crystal represents such a system. (For a rough understanding of this statement, it might be mentioned that an electron can also be regarded as a wave using the concepts of wave mechanics).

In metals, the order of magnitude of one electron per atom can move freely. Because of this very large number, the occupation of all the energy states available within this lattice is as dense as possible. If these energies of the electrons are plotted on a vertical scale (Figure 15), all the possible states are really occupied by electrons. If the horizontal axis is taken as any direction inside the crystal lattice, the band shown in Figure 15 indicates that the energy is not dependent on location, so describing the free mobility correctly. In such a representation, the energies of those electrons which remain bound to the atoms, forming their cores, can also be shown. In many cases it can be shown that these two groups of energy states, the "conduction band" and the "valence band" states, are separated by a "band gap". In the conductors, the width of this gap, giving the difference of energy between the highest valance state and the lowest in the conduction band, is much smaller than the thermal energy, or zero, or the bands even overlap.

Following this model, the situation in an insulator can be described by assuming a wide gap between the valence and the conductivity band, at any rate being larger than the thermal energy. This picture correctly describes the fact that there are no electrons available for conduction; the conduction band is empty.

In order to fit the semiconductors into this model, it is assumed that there are places within the crystal at which, locally confined, electrons are present which can relatively easily

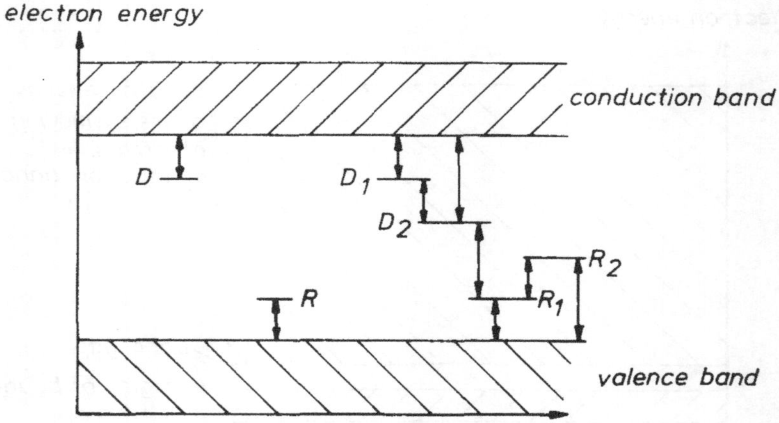

FIGURE 16. Energy bands in a solid with a band gap larger than the thermal energy with intermediate states; D = donor states, R = receptor states.

reach the conduction band. If the band gap in Figure 16 is assumed to represent an energy larger than the thermal energy, the material concerned will be an insulator. If, however, there are states like those indicated by D, having an (energy) distance to the conduction band, just close to the thermal energy, the electron can be lifted into the conduction band if it is hit by a larger thermal energy unit (phonon), thus increasing the conductivity of the crystal. A site such as D is called a "donor". Instead of being thermally activated by a phonon, the electron can also absorb a quantum of light, a photon. If the electron has reached the conduction band, the site D is lacking an electron. If an electron is coming close to it, it might be "attracted" by this availability of a state of lower energy, so the site D acts as a "receptor". There is also the possibility of a site, closer (energetically) to the valence band, being able to receive an electron from this band shown as R in Figure 16 ("receptor" site). If an electron is lifted to a site like R, an electron will be missing in the valence band. To a missing electron a positive electrical charge can be ascribed, called a "hole". If a receptor site like R is occupied by an electron and if a hole is coming close to it (in which picture it is assumed that holes can move in the valence band), the two can recombine, with the receptor acting as a donor under these conditions.

In natural semiconductors many more states are usually to be found as indicated by D_1, D_2, R_1, and R_2, with a large variety of possible exchanges. Today's semiconductor science uses these possibilities for tailor-made materials for many purposes.

Any interface within a crystalline structure and any surface can be represented electrically by a step potential preventing the electrons of the conduction band to pass or to leave it freely. The height of this step potential depends on the electronic state of the bulk material, the condition of the surface concerned, and the materials in contact with it.

We are just about to gather the first more detailed information about the influences of such surface potentials and the electronic states concerned with the different aspects of biocompatibility.

III. METALS

Metals are distinguished from other solids by a particular combination of properties: their high electrical and heat conductivity, their reflectivity, and their plasticity which allows them to be shaped by forging or cold working.

The crystal structures of the metals are mainly determined by the particular kind of bonding present in metals and the sizes of the atoms concerned. This allows for a high degree of miscibility of the metals and alloy formation. Thus, a large variety of combinations of properties can be realized.

A. The Characteristic Properties of Metals

The basic properties of metals can all be traced down to the few main features determining the bonding of the atoms and the crystal formation: the type of bonding results in the availability of the cloud of freely movable electrons and also in the kind of lattices formed. This will be discussed qualitatively only, but in a little more detail in the following subsections.

1. Electric Conductivity

The most characteristic property of the metals is their high electrical conductivity. This property is inherently correlated to the kind in which the atoms are bound: for the homopolar molecules, it had been mentioned in Section II.B that they can be regarded as having their valence electrons shared by all of the atoms concerned, the "cores". In this model the valence electrons have lost their individual attachment to the single atom, all of them are shared by all the atoms concerned, and, in this state, these electrons are providing the forces which overcompensate the electrostatic repulsion of the positively charged "cores". These forces can be understood and to some degree of accuracy can also be calculated using the calculus of quantum and wave mechanics. This kind of bonding is also called "metallic bond". The extrapolation of the homologous to the metallic bond used here may be felt to be somewhat unusual, but it appears to be a reasonable tool for explaining essential features starting from concepts more familiar to the reader. It has also been mentioned previously that solids can, to some approximation, be regarded as giant molecules. So, if more and more atoms are added and bound in the same manner, each of them contributing a portion of its valence electrons, a cloud of unlocalized electrons comes into existence. Because the electrons are not localized, they can be moved through the lattice by very small forces already. This results in the high electrical conductivity which can easily be measured.

2. Heat Conductivity

The existence of the large number of freely moving electrons does also account for another highly characteristic property of the metals, their higher thermal conductivity: if the average energy of the electrons is increased along one portion of the surface of a metal, this "disturbance" of the equilibrium state (the state of equal distribution of energy within the whole solid) is quickly passed on to neighboring regions and so carried on further because of the free mobility of the electrons.

3. Reflectivity of Metals

The particular type in which clean metal surfaces reflect the light is a direct consequence of the presence of the large number of free electrons. The surface electrons can follow the electrical field of the incident light for many frequencies very closely. Hence all wave lengths (colors in the visible region) which are strongly absorbed are also strongly reflected. Thus, the colors of such metals by transparent and by reflected light are complementary. A thin film of gold, for example, appears blue by transmitted light if illuminated by white light. Thus, these three characteristic properties of metals, the electrical and the thermal conductivity and the reflectivity, are based on the same feature, the presence of the large amount of not localized, freely movable electrons, the highly filled conduction band.

4. Plasticity of Metals and Work Hardening

There is another set of properties which can also be traced back to a single feature more or less common to all metals: the freedom with which the electrons can move within the lattice formed by the "cores" of the atoms concerned is to some extent supplemented by a

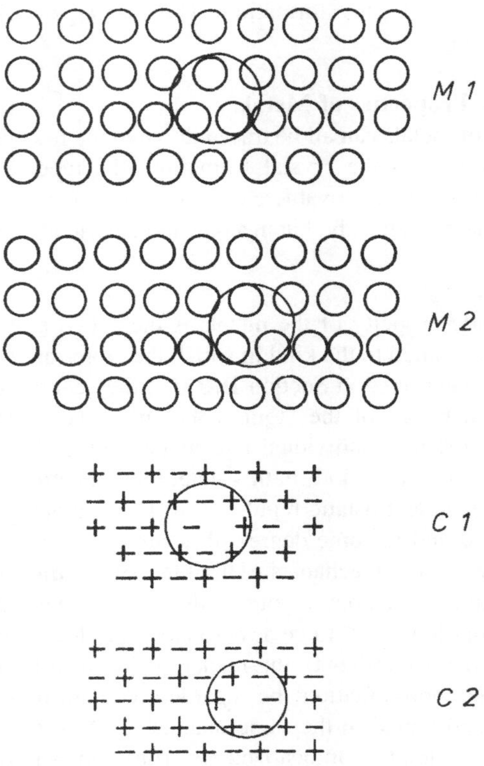

FIGURE 17. Schematic representation of a dislocation movement in a metal (M1 to M2) and in an ionic crystal (C1 to C2).

relative ease with which these cores can be moved. To qualitatively understand this ability, the schematic representation of a lattice plane in a metal and in an ionic crystal of Figure 17 can be used: the blocks M 1 and M 2 are supposed to represent lattice planes in a metal before and after the move of a vacancy for a unit step; the blocks C 1 and C 2 represent the lattice of an ionic crystal in a similar manner. In comparing the degree of rearrangement necessary for a change of the metal vacancy with the one necessary to move the vacancy in the ionic crystal and maintaining the electrical equilibrium, it can at once be realized that the degree of rearrangement is considerably higher in the second case. So, self-diffusion (and diffusion generally) must be expected to be lower in ionic crystals than in metals. This has some basic importance for the corrosion and, therefore, the biocompatibility aspect.

There are particular kinds of ordered arrangements of lattice defects which have already been mentioned previously and which have a basic importance in the understanding of the plasticity of solids. They are called ''dislocations'' and have already been shown in Figure 7. In Figure 17 dislocations are shown in a metal (as in Figure 7) and in an ionic crystal, together with a rearrangement necessary to move a dislocation for one unit step. For diffusion, the degree of rearrangement to achieve one unit step is higher in an ionic crystal than in a metal. Only if the degree of disorder of an ionic crystal is high enough (high temperature) does the energy necessary for such a rearrangement come close enough to the other energies concerned (maintaining the equilibrium) to effect this change. So, the plasticity of the metals is inherently based on or correlated to the particular type of homopolar or metallic bond.

However, this plasticity can be a temporary ability only. In Figure 8 the upper part shows the original position of the dislocation. If forces are acting as indicated by the arrows, the

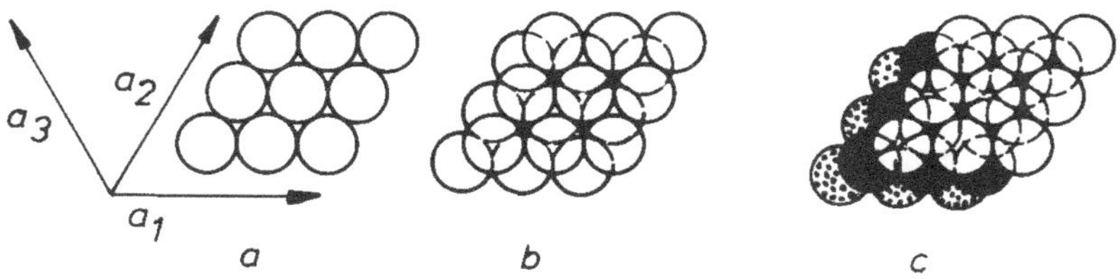

FIGURE 18. Close-packed structure of equally sized balls. (a) Monolayer and three axis in this basal plain; (b) second layer on top of the first with triangular interstitials on top of the balls of the first layer and above its interstitials. If the balls in a third layer are placed in those interstitials above the first layer, a 12121... sequence results; (c) stacking arrangement with 1231231... sequence.

dislocation will move through the sample, allowing for a change of the sample's contour. However, by performing this, the dislocation is annihilated. Thus, if the forces are going to act again, the sample will no longer be plastically deformable, it will have lost this ability. The same can happen if the dislocation, instead of being annihilated, is blocked by an obstacle like an inclusion or a precipitate. Groups of dislocations moving along different lattice planes can interact and, thereby, also block each other.

All of us have been using this "work hardening" of metals again and again: if a wire shall be separated in two pieces, it is commonly bent several times. In doing so, the wire becomes stiffer and stiffer in the bent region until it finally breaks. In many cases this hardening can be removed by heating the piece concerned to a temperature at which new dislocations are formed and/or the obstacles by which they have been blocked are annihilated.

B. The Main Phases of Metals

The forces binding the metal atoms to each other are without any directional preference. So, the description of the atoms within the lattice as balls offers a realistic approximation if there is some room for additional influences which might be attributed to the electrostatic repulsion of the positively charged "cores" of the atoms after they have contributed one of their electrons (or a part of it) to the "electron cloud", thus accounting for some deviations from the very close-packed arrangements. However, the two close-packed structures can be considered as the basic ones.

There are two ways by which close-packed layers of identical balls can be stored. Figure 18a shows such a close-packed layer in which each ball is surrounded by six neighbors in a hexagonal manner. If, to begin with, one ball is placed on top of this layer, it will find a position in the space at the center above three balls of the first layer. If, then, more balls are placed around the first one in the second layer, this second layer will, in itself, be identical with the first one, but shifted by one ball radius along two (a_1, a_2) hexagonal axes (in Figure 18b). For the third layer, there are two alternatives: its balls can be placed so that each of them is exactly above the ball of the first layer. This can be described by a sequence of layer 12121... It is called the hexagonal close-packed (hcp) structure because the simplest way of describing this arrangement is by using the three axes (a_1, a_2, a_3) shown in Figure 18a and the one perpendicular to them. The other way of arranging the balls within the third layer is by shifting them again by one radius further parallel to the a_1 and a_2 axis (Figure 18c). The position of the balls will not be identical to the two lower layers. Only if more layers are added can the sequence be repeated and will then be 12312312... The most simple way of describing this arrangement is by the face centered cubic (fcc) lattice (Figure 19) in which the body diagonal of the cube is chosen perpendicular to the planes in which the balls are placed in the above description (hexagonal planes).

In both these close-packed structures each ball is surrounded by 12 nearest neighbors.

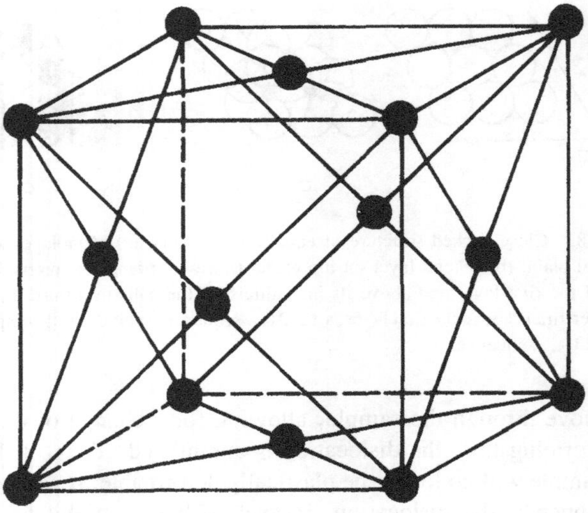

FIGURE 19. Face centered cubic lattice (fcc). It has four atoms or ions per unit cell.

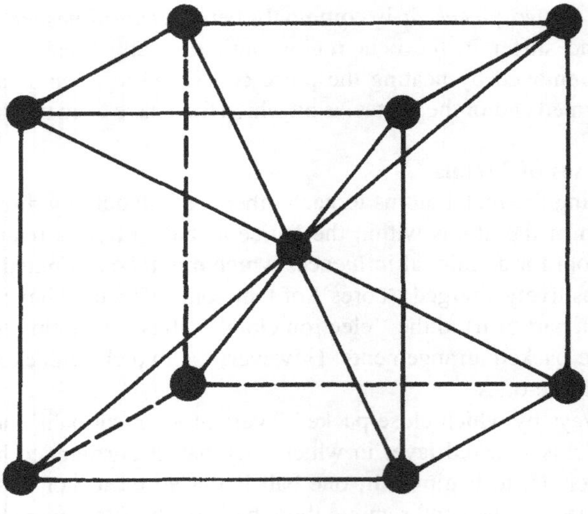

FIGURE 20. Body centered cubic lattice (bcc). It has two atoms per unit cell.

The space between the (rigid) balls occupies 26% of the total volume. There are two kinds of interstitials between the balls, one with a tetrahedral surrounding of the adjacent spheres and the other with an octahedral. There are one octahedral and two tetrahedral interstitials per ball. Another lattice often found in metals is the body centered cubic (bcc) structure. Here, the density is lower and each atom has eight nearest neighbors (Figure 20).

Most of the metals and alloys used in medical applications have structures which can be described with one of these three lattices. More details are given in Table 3.

The description of two of the most frequently occurring phases of metals, the close-packed structures, has shown how closely related these lattices are. So, if looked at from the point of view of the fcc structure, any deviation from the stacking sequence ... 123123 ... to, for example, ... 12312123 ... can be regarded as a lattice defect, a stacking fault. The same

Table 3
CRYSTALLOGRAPHIC DATA OF SOME PURE METALS

Metal	Phase	Type of lattice	Dimensions of unit cell (10^{-10} m)	Stable at temp. (°C)
Iron	α-Fe	bcc	a = 2.8606	
Cobalt	α-Co	hcp	a = 2.514 c/a = 1.633	Below 450
Nickel	Ni	fcc	a = 3.5168	
Gold	Au	fcc	a = 4.0704	
Silver	Ag	fcc	a = 4.0778	
Copper	Cu	fcc	a = 3.6080	
Molybdenum	Mo	bcc	a = 3.1410	
Aluminum	Al	fcc	a = 4.0414	
Titanium	α-Ti	hcp	a = 2.95 c/a = 1.60	Below 882
Tantalum	Ta	bcc	a = 3.298	
Platinum	Pt	fcc	a = 3.9160	
Tungsten	W	bcc	a = 3.1584	
Chromium	Cr	bcc	a = 2.879	

applies if looked at from the point of view of the hcp structure, so that an arrangement like ... 1212312 ... can also be regarded as a stacking fault within the hexagonal sequence. Energetically, these two structures are also not far apart for most of the metals.

Qualitatively, this close similarity of the structures concerned and the low energy barrier separating them allows one to understand the relative ease with which a metal can assume different phases. In many cases (but not always) the phase into which the liquid metal solidifies is called the α-phase, those into which it transforms with further decreasing temperature are called β-, γ-, etc. phase.

Not all phase transformations affect the whole body concerned. Because of the energies involved, a given crystal is often transformed incompletely so that two phases can exist side by side, one of which being the true equilibrium phase at the temperature concerned; the other being a metastable state.

This possibility of creating metastable states is of great technical importance. If, for instance by rapid quenching, one phase is distributed finely dispersed within another, the matrix phase, the fine particles can block the movement of dislocations within the matrix phase. Thus, the mechanical behavior of the metal is changed, e.g., its fatigue properties. By increasing the temperature again to the transformation temperature, the dispersed phase can be dissolved and/or its distribution within the matrix phase can be changed completely by choosing a different cooling process. So, the mechanical properties of the same metal can be changed severely depending on the thermal history of the particular piece.

C. Alloys

Any mixture of two or more metals is called an alloy. The close similarity of the lattices into which most of the metals crystallize also accounts for the relatively good miscibility of many metals allowing for many solid solutions of one metal in the other.

The different phases occurring in a mixture of two metals and their distribution as a function of concentration and temperature is represented graphically in the "phase diagrams". In these diagrams the vertical axis represents the temperature; the horizontal axis represents the concentration of the metals of this "binary alloy". These representations can

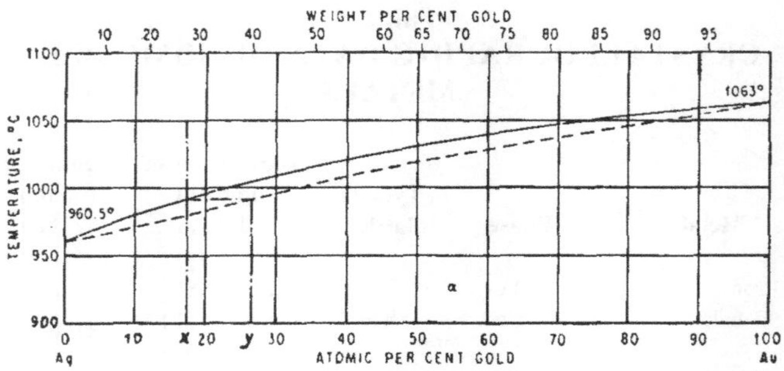

FIGURE 21. The Ag-Au phase diagram as an example of complete miscibility. The width of the gap between the liquidus (upper curve) and the solidus (lower curve) is mainly caused experimentally. Theoretically, it should be 1.3°C only. The whole system is a continuous series of solid solutions, completely free from any ordering of the Ag and Au atoms on the lattice sites of the fcc lattice at temperatures between the solidus and room temperature. However, there are indications of some short-range orders.[22]

be regarded as a projection of the complete four-dimensional system considering all four thermodynamically relevant variables (pressure, p; volume, v; temperature, t; and concentration, c) onto the c,T plane. The lines drawn within these diagrams represent boundaries along which two phases are in equilibrium with each other. Within the areas separated by these lines, either one or two phases can exist.

The most simple case of a phase diagram is shown in Figure 21 describing the gold silver alloys. Starting at the Ag side at high temperatures, pure silver will be found to solidify at a well-defined temperature, T_s = 960.5°C. If something is added to a pure metal that forms a solid solution, the solidification temperature is either lowered or raised if the melt is in equilibrium with a mixed crystal containing either less or more of the addition, respectively. If a melt of composition X in Figure 21 is cooled, a temperature will be reached at which a crystal containing gold of the concentration Y can exist. Thus, below the "liquidus temperature" two phases will exist side by side until, on further cooling, all of the melt will have solidified, thus defining the "solidus temperature". Below this temperature the atoms of the two metals are distributed purely statistically on the lattice sites. Any deviation from such a purely statistical distribution within a solid solution is called "super structure". The gold platinum system of Figure 22 also shows a complete solid solution behavior just below the solidus curve. However, this solubility of the two boundary phases into each other is not complete. At lower temperatures, the solubility is limited and there is a region in which the two phases exist side by side. The diagram also gives some indication for the existance of an "intermetallic compound" or "intermediate phase" ($PtAu_3$).

The iron-chromium phase diagram (Figure 23) again shows a wide range of miscibility interrupted only by the intermediate σ-phase. However, it has another peculiarity: if pure iron is cooled from its molten state, it solidifies into a body centered cubic lattice, called α-phase. In cooling further it undergoes a solid state phase transition into the face centered γ-phase at 1390°C. This phase is maintained on further cooling until 910°C. At that temperature the body centered α-phase is resumed again. At a still lower temperature, iron undergoes another kind of transition from the paramagnetic into the ferromagnetic state. If another component is added to a metal undergoing solid state phase transitions, of course, these different phases of the one metal will extend into the alloy region and give rise to more complicated phase diagrams. The extension of the γ-phase of the iron leads to the γ-loop in the alloy diagram.

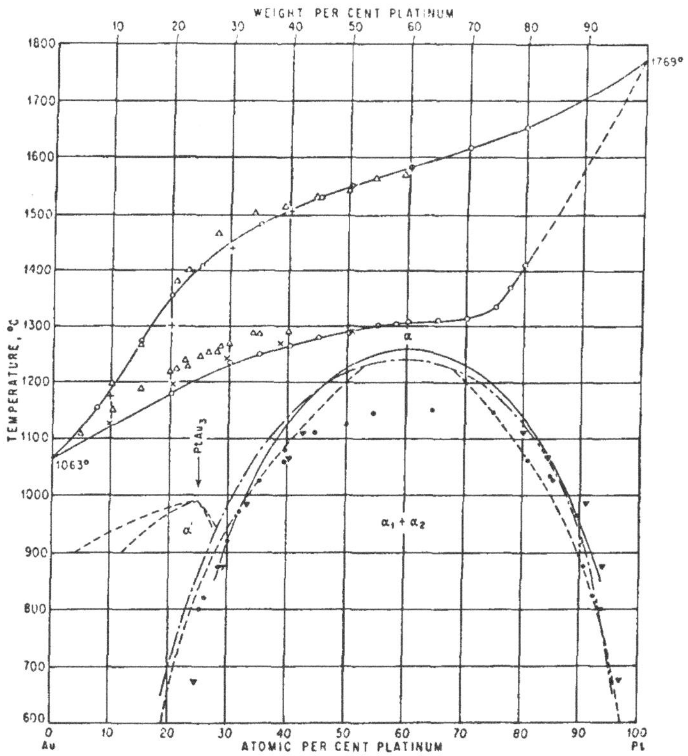

FIGURE 22. The Au-Pt phase diagram has a large and irregularly shaped gap between the liquidus and the solidus. Immediately below the solidus curve there exists a continuous range of solid solutions, while at lower temperatures, a "two phase region" can be seen. Its existence is largely responsible for the mechanical properties of these alloys.[22]

The cobalt-chromium system shown in Figure 24 contains an "eutectic decomposition". It also contains two intermediate phases, ϵ and σ (in the areas not containing any inserts the two boundary phases coexists). Another peculiarity can be seen in the representation of the cobalt-molybdenum phase diagram of Figure 25. Here the intermetallic compound $MoCo_3$ is formed by a peritectoid reaction at 1020°C and so is the Mo_6Co_7 phase at 1500°C.

Figure 26 gives one of the other "boundary" binary alloys from which the basic ternary alloys of the Co-Cr-Mo systems are derived which have gained so much importance in different fields of orthopedic surgery and dental implantology. Titanium alloys have gained more and more importance during recent years because of their excellent biological compatibility and high mechanical strength.[43] Originally, these alloys had been developed for aero- and space applications with vanadium as one of the alloying metals. Because of the toxicity of vanadium, a Ti-Al-Fe alloy was developed. Figures 27 and 28 show two of the binary alloy systems from which this ternary system is derived.[44]

The representation of alloys with three or more metals cannot be given in such relatively simple two-dimensional diagrams. For alloys containing three metals, the triangular representation can be used as shown in Figure 29. The corners of the triangle represent the pure metals (A, B, C), while the sides give the concentration variation of the binary alloys concerned (AB, BC, AC). For any position inside the triangle, the concentrations $c_ac_bc_c$ can be given considering that $c_a + c_b + c_c = 1$. Inside such a triangle the phase boundaries can be given at one particular temperature only. If such graphs are made for different temperatures and arranged on top of each other, the equilibrium phase distribution of a

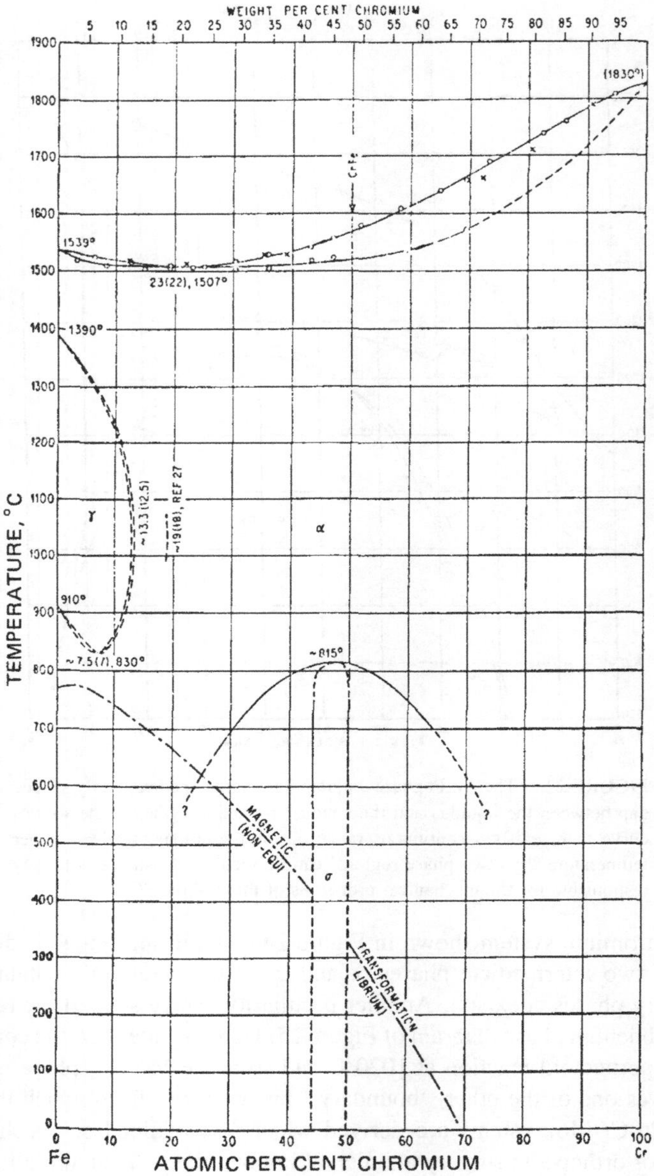

FIGURE 23. The Fe-Cr system is of basic importance for the stainless steels. Besides the wide range of complete miscibility between 1500 and about 800°C, it shows the α-phase loop on the iron rich side, the boundaries of which are difficult to reproduce because of their sensitivity to impurities. The boundaries of the α-phase, an "intermediate phase" in the equiatomic region, can hardly be determined exactly because of an extremely low rate of formation. It has a lattice of tetragonal symmetry with 30 atoms in the unit cell. Because iron is ferromagnetic below 770°C ("ferromagnetic Curie temperature"), this "magnetic transition" extends into the solid solution region continuously.[22]

ternary system can be described. However, of course, such a three-dimensional representation (Figure 30) can be used for demonstration only. For quantitative evaluations of ternary systems, mathematical methods must be used.

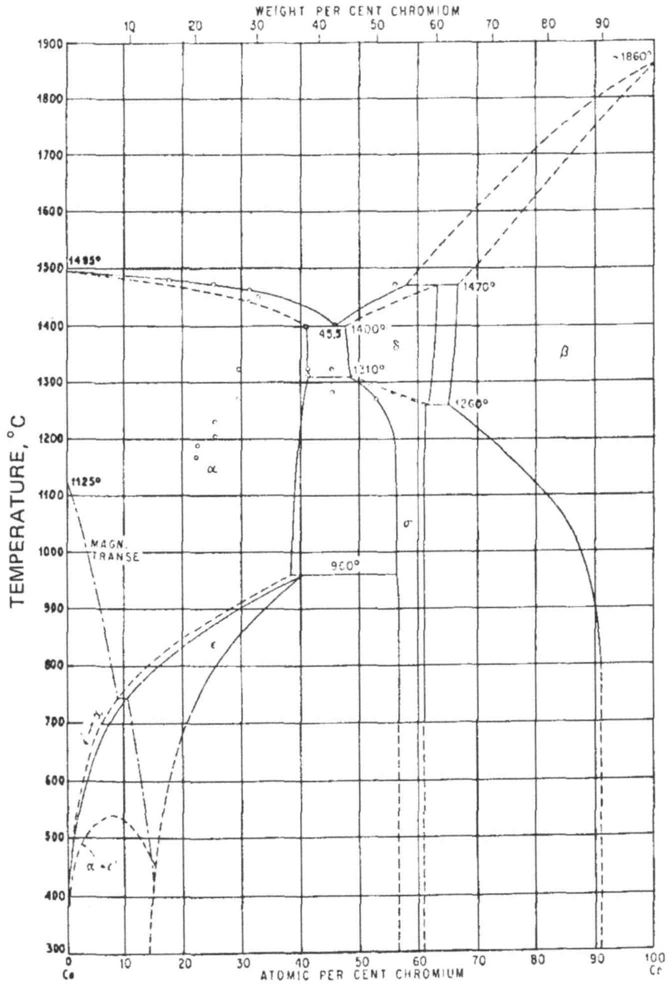

FIGURE 24. The Cr-Co system contains a "eutectic decomposition" at 1400°C into the α- and σ-phases. It is characterized by the existence of two intermediate phases, a phase of singular composition Cr-Co (ϵ) formed by a peritectoid reaction of α- and σ-phase at about 1260°C, and a phase of variable composition (σ) with 55 to 62 wt% Cr having a tetragonal unit cell containing 16 Cr and 14 Co atoms.[22]

IV. CERAMICS

From the point of view of materials' science, all nonmetallic and nonorganic materials belong to the ceramics. There is also the other meaning of the word ceramics which describes the art of pottery. However, most of the ceramics which have been considered as implant materials since about 1964 and those which have found their clinical application during the last decade have a very faint relationship only to the silicates of the classical ceramics. Most of them belong to the group of oxide ceramics. Technologically, these oxide ceramics have much more similarity with powder metallurgy. This is also true if looked at from the structural point of view.

Figure 31 gives a schematic survey. All of the classical ceramics contain silicates, at least a portion of which is introduced by the use of clays, thus allowing the shaping of the desired

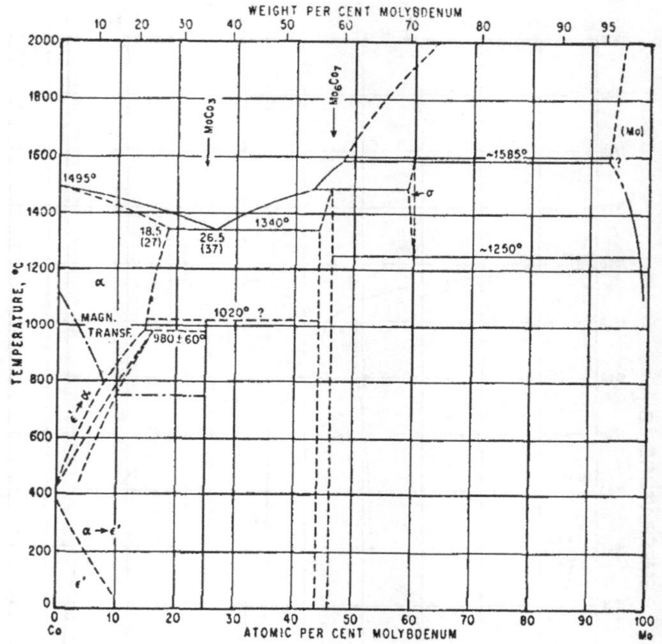

FIGURE 25. The Co-Mo phase diagram also contains a "eutectic" (at 1340°C and 26.5 atomic % Mo) and compound $MoCo_3$ formed by a peritectoid reaction at 1020°C of α-Co with the Mo_6Co_7 phase. This Mo_6Co_7 phase is formed peritectically at about 1500°C.[22]

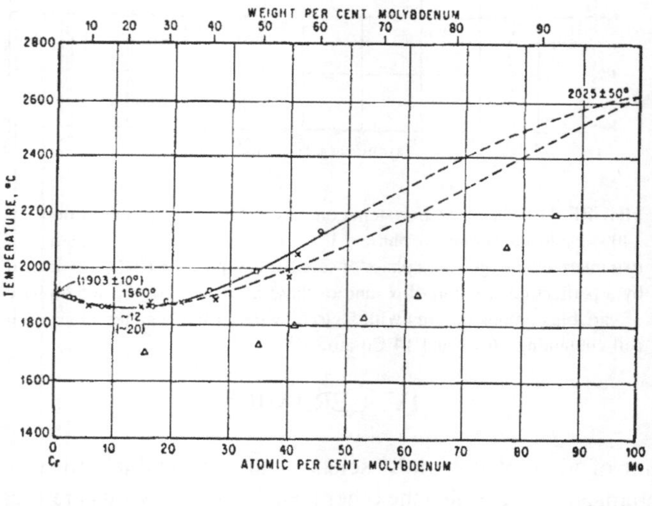

FIGURE 26. The Cr-Mo phase diagram shows one of the systems of which the Co-Cr-Mo alloys consists. It is another example of a continuous, one-phase series of solid solutions.[22]

parts by mixing with the correct amount of water without further additives. The oxide ceramics start from industrially prepared, mostly pure powders to which some plastics are added temporarily to allow for an approximate shaping. In some of the ceramics which are prepared from industrially made raw materials, however, the final compound is sometimes not formed until the firing process. The glasses and glass ceramics are, of course, also made

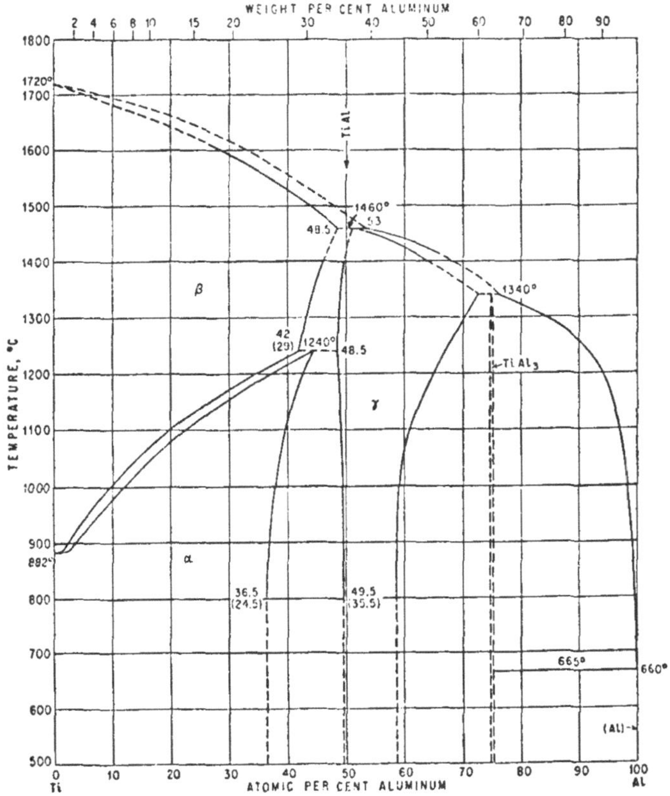

FIGURE 27. The Ti-Al alloy system has a rather wide range of solubility of Al in α-Ti. It also contains the two compounds TiAl and the $TiAl_3$. The high-strength Ti alloys used for orthopedic implants are located in the α-Ti area at about 5 wt% Al.[22]

from silicates, from which a melt is made out of the parts formed or preformed by solidification. The carbons are either made from carbon powders or from plastics which are converted into carbons by a special heat treatment.

In all of the ceramics, the shaped or molded raw piece is transformed into a solid by a process called "firing". However, there is a major difference: in the silicate containing ceramics one or several melts are formed during the firing cycle which solidify and partially crystallize during the process. Thus, the final product always is a mixture of different phases. In oxide ceramics, the shaped agglomeration of powder particles is transformed into a polycrystalline solid by solid-state reactions only, mostly by diffusion, and if pure powders are used, the final piece is a one-phase only material, the properties of which fully control the property combination of the product.

A. Oxide Ceramics

Most of the ceramics used for implant purposes belong to the group of oxide ceramics. Therefore, their fundamentals, as compared to other solids, will be discussed in some more detail.

In other chapters homopolar and heteropolar compounds have been distinguished either by sharing some of their electrons with no close localization with one or the other atom or by having them rather clearly allocated to the atoms concerned, thus rendering them clearly

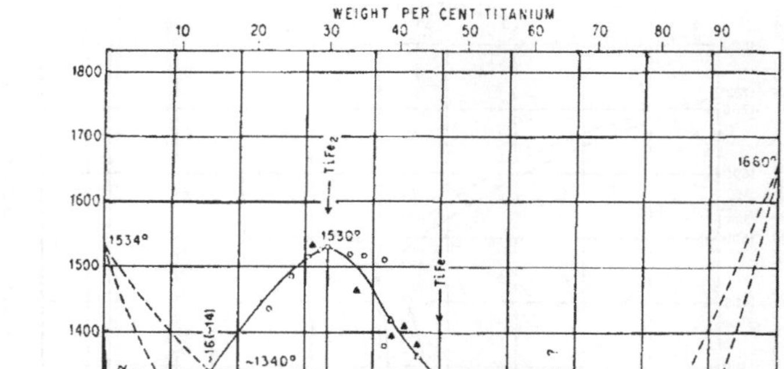

FIGURE 28. The Ti-rich region of the Ti-Fe system is a portion of the ternary alloy TeAlFe recently developed for orthopedic applications. It shows a eutectoid decomposition of the β-Ti phase at about 590°C (more recently corrected to about 625°C) and a very much narrowed field of existence of the α-Ti phase as compared to the Ti-Al system (see Figure 27).[22]

into ions. In metals, the freely moving electrons have been found to be an essential feature controlling most of their basic properties.

In the ceramics and particularly pronounced in all oxide ceramics, the ionic nature of the bond is predominant. This immediately implies that ceramics always contain more than one kind of atom, with carbon being the only exception. Oxide ceramics usually consist of one or a very few metal oxides.

The metals from which oxide ceramics are known belong to the group of the least noble metals. These oxides are already very stable compounds (with a high energy of oxide formation). However, they must also have a crystal phase which is stable at all temperatures concerned.

The crystal structures of the oxide ceramics are mainly determined by a close-packed arrangement of the oxygen ions into the interstitials of which the metal ions concerned are distributed. The details of this distribution are mainly controlled by the requirement of electrical neutrality and the size of the ions.

Some basic features of the close-packed structures had already been mentioned in Section

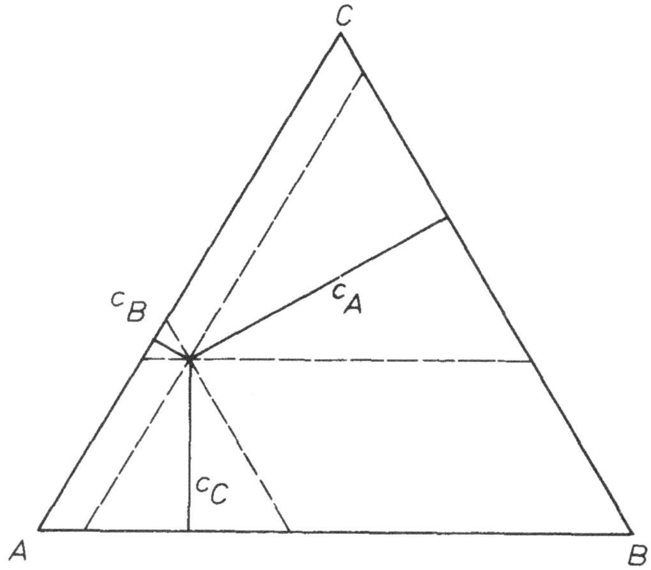

FIGURE 29. Schematic representation of a ternary system giving the concentrations of the components c_A, c_B, and c_C graphically.

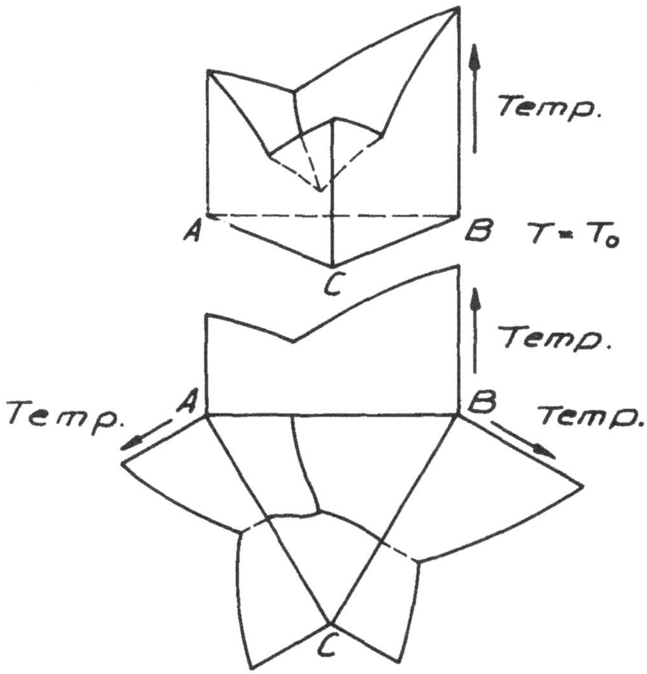

FIGURE 30. Perspective view of a three-dimensional ternary phase diagram (upper portion) and an "unfolded" view showing the three "boundary" binary phase diagrams (below).

III.B. The main representative of the oxide ceramics, alumina or corundum ceramic, has a hexagonal close-packed arrangement of the oxygen ions. The Al^{3+} ions occupy two thirds of the octahedral interstitials. Each aluminum ion is surrounded by six oxygen ions, three

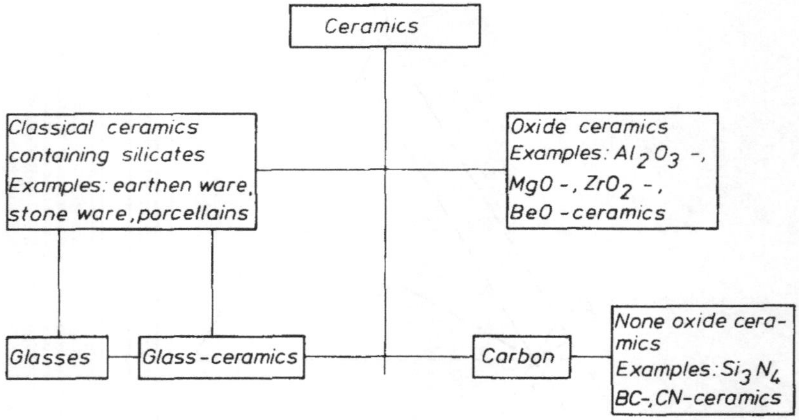

FIGURE 31. Ceramic materials.

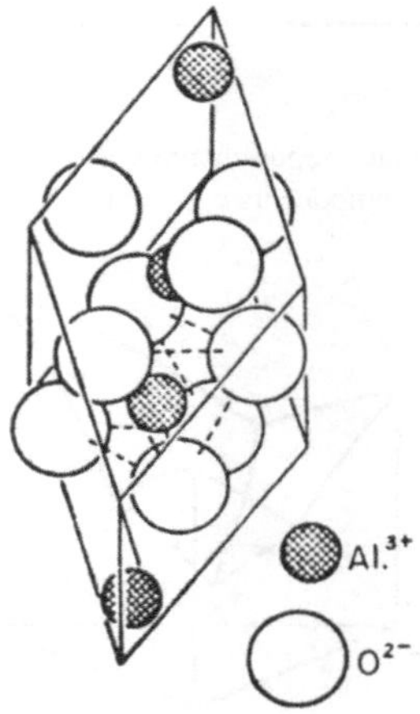

FIGURE 32. The ion arrangement in the corrundum lattice.

of which form a regular triangle on one side, the other three form a similar triangle on the other, with the two planes of the triangles being parallel and the triangles being twisted by 180°.

Figure 32 shows the arrangement of the ions in the corundum lattice schematically. The unit cell (space group $D^6{}_{Ba}$) contains two Al_2O_3 "molecules" with a distance between the oxygen ions of 2.495 Å and a ratio of the dimension of the unit cell parallel to the hexagonal axis (c) to its dimension perpendicular to the hexagonal axis (a) of 1.366. This corundum or α-alumina phase is crystallographically identical with ruby and sapphire. Corundum is the colorless version of which, for example, scratch-proof "glasses" of more expensive wrist watches are made. The other two are distinguished by very small amounts of some dopants like iron, cobalt, and/or manganese influencing the color only. The α-phase is the most stable modification of alumina. If an Al_2O_3 melt is cooled below 2050°C it solidifies into the α-phase which is maintained on further cooling without any phase transition. If, however, aluminum hydroxides, e.g., resulting from the Bayer process, are heated to several hundred degrees, but kept below 1100°C, other, less stable phases are formed, of which γ-alumina is the most important. It has a cubic face centered arrangement with a lattice constant of 7.90 Å. However, all these less stable phases of alumina transform into the α-phase on

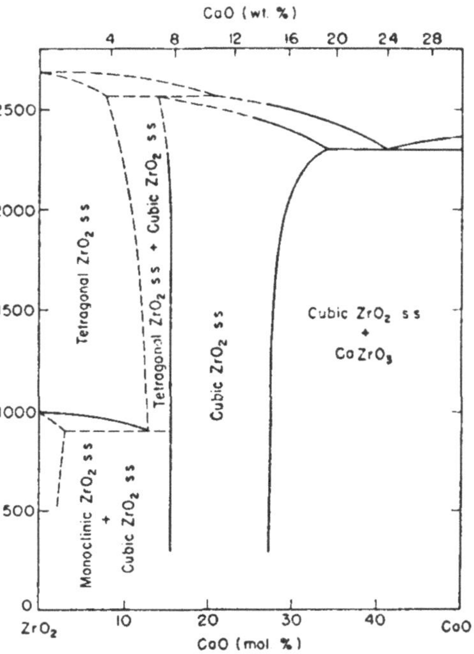

FIGURE 33. The ZrO_2-CaO phase diagram.

heating to about 1100 to 1500°C, this transition being irreversible under normal conditions. If, however, an α-alumina powder is heated above the melting temperature (2050°C) and rapidly quenched as in plasma spraying, a considerable portion of the powder particles resolidify in the γ-modification.

Another oxide ceramic, magnesia, crystallizes in the cubic face centered lattice (periclase). Its unit cell contains four MgO "molecules" and has a lattice constant of 4.20 Å. There is no other modification of MgO.

More recently a zirconia ceramic has been considered for medical applications (as a coating onto metal stems of hip replacements).[33] This material, ZrO_2, undergoes several phase transitions upon cooling from the molten state to room temperature. It is polymorphic. At high temperature and down to about 1000°C, the tetragonal phase is stable with the parameters, a = 5.07 Å with a calculated density of 6.10 g/cm^3. The low-temperature monoclinic modification (baddeleyite) has the parameters, a = 5.174 Å, c = 5.308 Å, with angle β = 80.8° and a calculated density of 5.56 g/cm^3. Thus, the phase transition at 1000°C is associated with an approximate 10% decrease of density or volume increase rendering it impossible to produce a sintered body from pure zirconia. If, however, some other oxides like calcia (CaO) or yttria (Y_2O_3) are added in amounts of 5 to 15 mol%, a cubic solid solution can be stable down to room temperature. The phase diagrams of the binary oxide systems concerned are shown in Figures 33 and 34. The relative amounts of the phases actually present at room temperature strongly depend on technological parameters because of the different kinetics of phase precipitation and formation. This material can be sintered to high density. It is particularly distinguished by an exceptionally high ionic conductivity of oxygen (above about 500°C), rendering it into a solid electrolyte for which more and more very interesting applications are considered and partially already realized in the field of energy conversion and control of combustion processes.

For the sake of completeness, two more simple oxides belonging to the group of oxide ceramics should be mentioned. Beryllia (BeO) has one stable hexagonal structure (wurtzite)

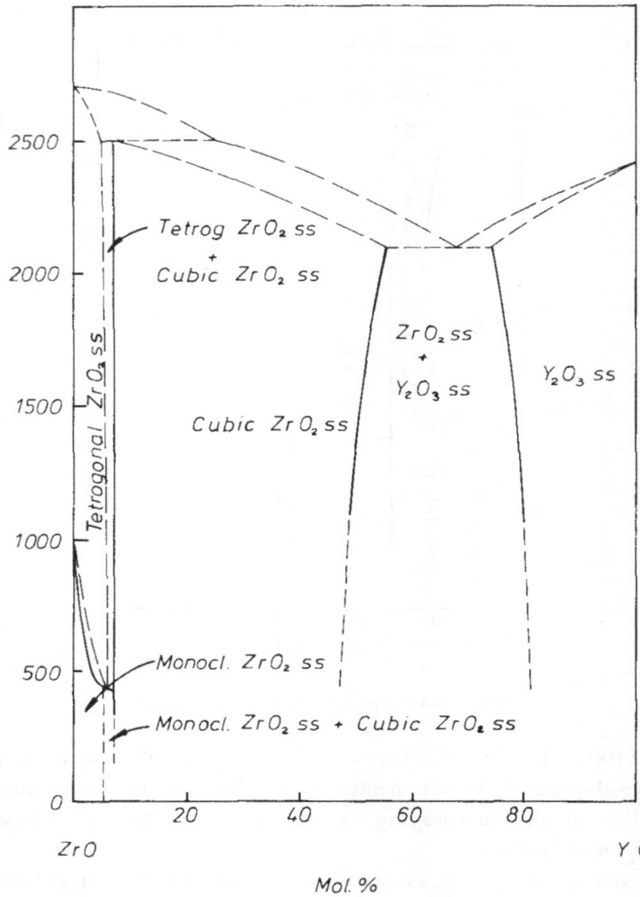

FIGURE 34. The ZrO_2-Y_2O_3 phase diagram.

with only a high ionic packing density (a = 2.70 Å; c = 4.39 Å; distance Be and O, 1.645 Å only). The usual density is 3.02 g/cm³. It is distinguished by a heat conductivity approaching that of metals, while its electrical conductivity is as low as with the best insulators. Thus, BeO ceramics are used as heat dissipating substrates for high-quality electronic circuits. However, the toxicity of BeO particles in the lung is a severe handicap of this material.

Thoria (ThO_2) too, has one stable phase only with a face centered cubic lattice (a = 5.59 Å) and a calculated density of 10.05 g/cm³. With a melting temperature of 3300°C (and a boiling temperature of 4250°C), it is one of the most stable oxides. In the times of gas light, thoria was used for incandescent Auerlight mantles. Today its applications are limited to crucibles for high-temperature treatments and similar tasks and to measuring extremely low oxygen partial pressures, e.g., in sodium of sodium-cooled nuclear reactors.

There is one mixed oxide also regarded as belonging to the oxide ceramics. $MgAl_2O_4$, spinel, has a cubic structure with a lattice constant of 8 Å, 8 "molecules" in the unit cell, and a density of 3.84 g/cm³. Figure 35 gives the reproduction of the spinel diagram in which it can be considered as a very stable interoxide compound. Table 4 summarizes the main features and properties of the oxide ceramics.

B. Nonoxide Ceramics

The main representatives of nonoxide ceramics considered for medical applications are

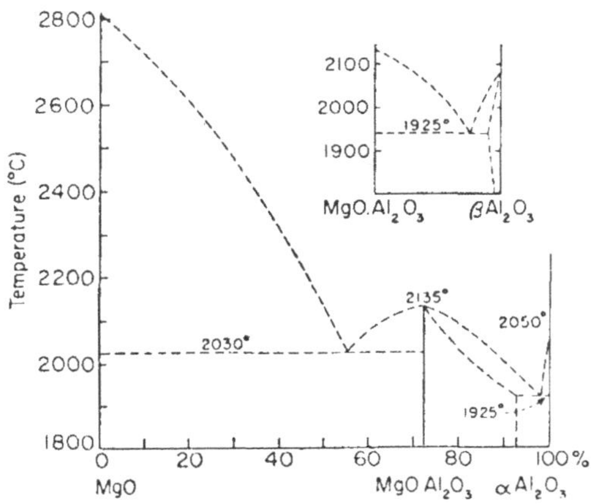

FIGURE 35. The Al_2O_3-MgO phase diagram.

the ceramics based on SiC, Si_3N_4, and carbon or combinations of them and with some oxides.

These materials have to some extent rather extreme properties because of the close-packed structures of their relatively small-sized ions. Technically their mechanical properties at high temperatures are of main importance; their mechanical strength and chemical inertness have been reasons for considering them as materials for medical applications.[16]

Silicon carbide (SiC) crystallizes in numerous hexagonal, rhomboidic, and cubic polymorphic lattices with an always identical unit cell dimension in the hexagonal a-direction of 3.078 Å and different dimensions in the c-direction depending on the sequence of the layers in c-direction. The basic structural unit is a tetrahedron in the center of which one of the two kinds of ions is located, while the other kind of ions is at the corners belonging to three other tetrahedrons simultaneously. The unit cell dimension of the only cubic modification is 4.358 Å. These different polymorphic phases are of some importance technologically, however, they hardly influence the properties of the SiC based materials.

Silicon nitride (Si_3N_4) exists in two hexagonal phases in which each silicon ion is tetrahedrally surrounded by four nitride ions, each of which belonging to three neighboring tetrahedras. These tetrahedras are arranged in layers with two different sequences: in the case of the β-Si_3N_4, the first, third, and fifth layers are identically arranged; in the α-phase, the first, fifth, and ninth are identically arranged. This latter lattice can contain considerable amounts of oxygen in solid solution. The lattice constants of the β- and the α-phase are a = 7.608 and c = 2.91 Å and a = 7.753 and c = 5.62 Å, respectively.

Sialon ceramics are based on the β-Si_3N_4 phase with relatively large amounts of aluminum oxide in solid solution resulting in an increase of the unit cell dimensions. The main properties of the different grades of SiC and Si_3N_4 ceramics are given in Table 5.

C. Carbons

Since about the late 1960s, carbons have been considered for prosthetic devices.[3,6] There are two main types of crystalline arrangements of carbon atoms: the graphite type and the diamond type. In the graphite structure, the atoms are forming layers composed of hexagons with relatively high binding forces within the layers and a weak bonding between adjacent layers. There are two different stacking sequences of the layers shown in Figure 36. The 12121... sequence resulting in a hexagonal symmetry with a = 2.46 and c = 6.708 Å (2-H graphite) and the 1231231... sequence with a rhomboidic symmetry and a = 2.46 and c = 10,062 Å (3-R graphite). The other main type of arrangement of the carbon atoms is

Table 4
BASIC PROPERTIES OF OXIDE CERAMICS (ROOM TEMPERATURE OR CLOSE TO ROOM TEMPERATURE VALUES)

Ceramic	Main crystal phase	Density (g/cm^3)	Young's modulus (N/mm^2)	Compressive strength (N/mm^2)	Bend strength (N/mm^2)	Hardness (HV) (N/mm^2)	Specific heat (Ws/kg)	Heat conductivity (Ws/K cm s)	Electrical conductivity ($\Omega^{-1}cm^{-1}$)	Thermal expansion ($10^{-6}/K$)	Melting temp (°C)
Al_2O_3	Corundum	4	$0.38 \cdot 10^6$	4000	400	23,000	0.862	0.192	$<10^{-15}$	6	2050
MgO	Periclase	3.57	$0.09 \cdot 10^6$	160	130		0.982	0.368	10^{-16}	11	2800
ZrO_2	Baddeleyite	6.10	$0.17 \cdot 10^6$ [a]	2100[a]	150[a]	16,000[a]	0.526	0.017	10^{-8} [a]	8—12[a]	2715[a]
BeO	Wurtzite	3.02	$0.31 \cdot 10^6$	1400	200	1,500	0.035	2.645	$<10^{-16}$	5	2650
ThO_2	Fluorite	10.05	$0.14 \cdot 10^6$	1500		80	0.228	0.123	$\sim 10^{-12}$	9.6	3300
$MgAl_2O_4$	Spinel	3.60	$0.23 \cdot 10^6$	1000	150	2,070	0.806	0.168	$\sim 10^{-12}$	6.7	2135

[a] Depending on the kind and degree of stabilizing dopant.

Table 5
PROPERTIES OF DIFFERENT GRADES OF SiC AND Si_3N_4 CERAMICS[a]

Product	Density (g/cm^3)	Porosity (%)	Bend Strength (N/mm^2) (20°C)	(1000°C)	(1400°C)	Young's modulus (10^3 N/mm^2)
Si_3N_4 sintered	3.3	<2	500	450	300	310
Si_3N_4 hot pressed	3.2	0	700	650	400	320
SiC sintered	3.15	<2	400	450	400	400
SiC hot pressed	3.2	0	550	550	450	420

[a] Ceranox of Annawerk, Rödental, Germany.

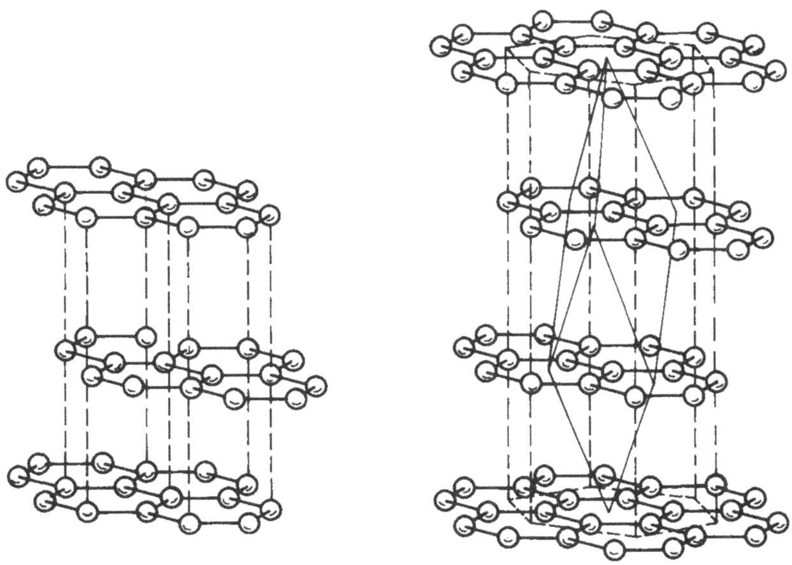

FIGURE 36. The graphite structures 2-H (left) and 3-R (right).

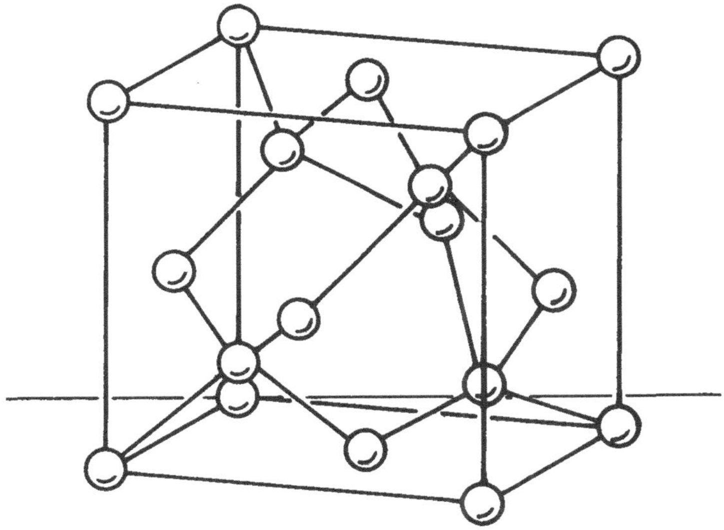

FIGURE 37. The diamond structure.

the diamond structure shown in Figure 37. It has a cubic symmetry with a = 3.559 Å and 8 atoms per unit cell.

Besides the crystalline phases, several more or less amorphous (with different degrees of short-range orders) versions have been realized during the last decades. These glassy or vitreous carbons are made from hydrocarbons by a special heat treatment. The largest applications of vitreous and pyrolytic (LTI) carbons (formed by chemical vapor deposition of hydrocarbons at high temperatures onto a carbon substrate) have been in artificial heart valves.[23]

For orthopedic applications, three different kinds of carbons and carbon-containing composite materials have been developed.[10,11,35]

Table 6
MECHANICAL PROPERTIES OF SOME CARBON BASED MATERIALS

Property	Isotropic carbon	SiC/C composite	C fiber reinforced carbon
Density, g/cm^3	1.8	2.6	1.7
Bend strength, N/mm^2 (static)	140	220	800
Young's modulus, K, N/mm^2	25	100	140
Porosity, %	10	0	7

1. High-strength isotropic carbon has a glassy carbon-like, but porous structure. It is hard, brittle, and has a relatively good wear resistance. It is produced by pyrolysis of a semicoke.
2. The silicon-carbide/carbide composites are manufactured by impregnating a basic porous carbon with liquid silicon at about 2000°C during which process the silicon is completely transformed into SiC. Thus, a composite results consisting of the two phases, SiC and C.
3. The carbon fiber reinforced carbons are made by impregnating the filaments (after orienting them in the desired direction or directions) with a pitch, compressing, heat-treating, and carbonizing or graphitizing.

The characteristic properties of three materials are given in Table 6.[10] The combination of one surface of the SiC/C composite against a surface of isotropic carbon has been shown to have the most favorable wear behavior within a particularly selected group of articulating surfaces.[30]

In evaluating the properties of these carbons for applications in the field of high load-bearing joints, it must be kept in mind that the high strength of the carbon fiber reinforced materials is available in the direction of the fibers only. As soon as the load direction deviates from the fiber axis, the static strength decreases considerably and the fatigue strength decreases even more. Thus, considering the multidirectional load situations under which all bone replacements must function, the available fatigue strengths of these materials are in the order of the pure and finely grained cobalt chromium molybdenum alloys. As far as the good frictional and wear properties of many graphites in mechanical engineering applications mentioned with the intention of recommending such materials for medical applications, it must be remembered that these favorable properties result from the layered structures of the graphites mentioned above. This allows for the sliding of the layers against each other until finally the top layers scale off the surface, and thus, inherent with this kind of friction, result in wear particles. Such wear debris can be dealt with in most technical applications, but biological systems lack a filtering and lubrication arrangement suited for these particles.

D. Glasses and Glass Ceramics

The glasses which have been considered for implant systems are based on silica, SiO_2. If multicomponent systems containing silica are cooled down from their truly liquid state (with low viscosity), they solidify without crystallization. This solidification is spread over a large temperature interval during which the viscosity increases by many orders of magnitude reaching a really solid, but still metastable state. At room temperature the mechanical properties of these glasses are very similar to many polycrystalline solids, but without any long-range ordering in the atomic arrangement.

However, these solidified melts have a marked tendency to form relatively large domains of short-range order. This tendency results from a particular property of the silicon oxygen

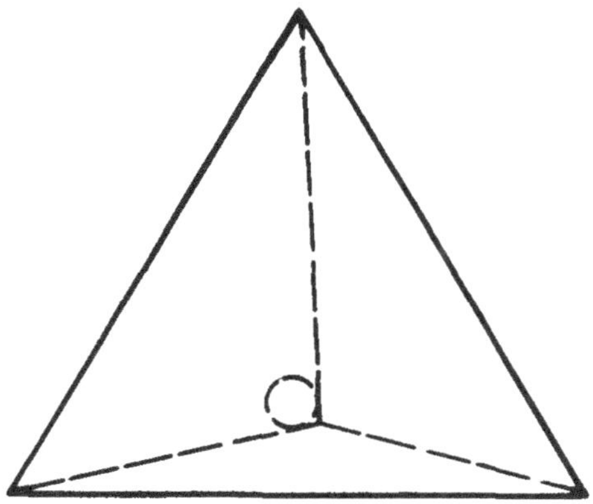

FIGURE 38. Tetrahedral. In the case of SiO_2, the center is occupied by a silicon ion; the corners are the centers of oxygen ions.

system which originates mainly from the high electrical charge of the Si^{4+} ion, its small size, and the ability of the silicon oxygen system to acquire different kinds of bonds. This results in the formation of a tetrahedral arrangement of the oxygen ions in the center of which the silicon ion is located, as shown in Figure 38. Because of the high electrical charge of the silicon ion, neighboring tetrahedrals are grouped sharing one oxygen ion only (largest distance between the Si^{4+} ions possible), resulting in configurations like those shown in Figure 39.

If pure SiO_2 is cooled down from its molten state (above 1725°C) slowly enough so that equilibrium conditions can be reached, it crystallizes in a hexagonal structure (cristobalite) in which the oxygen tetrahedrals containing the Si^{4+} ion (Figure 38) assume a three-dimensional, relatively loose arrangement.

Between 870 and 573°C, the high temperature modification of quartz is stable which is followed by the low temperature quartz, in which the tetrahedrals are packed more closely and in screw-type manner, allowing for left- or right-hand screw sequences. If a pure melt of SiO_2, however, is cooled faster, the above-mentioned equilibrium arrangements can only be achieved within small atomic distances involving a few of the tetrahedrals only, resulting in a macroscopically disordered structure, the quartz glass. This phenomenon can also be described in another way: because of the formation of the more or less complete chains of SiO_4 tetrahedrals within the low viscosity state already, the formation of large, well-ordered arrangements is much more unlikely than, for example, in a metal melt containing atoms only. (Atoms can be put together in any orientation, so that much more complex entities like groups of two or more tetrahedrals can be coupled to each other in particular orientations only.) Thus, the silica melts can reach enormously higher degrees of undercooling than metals or salts. This high nucleation barrier in a typical glass is important for the devitrification process to be discussed below.

If foreign atoms are added to the SiO_2 melt, the bonding between the oxygen tetrahedrals can be decreased, thus reducing the solidification temperature. This ability of the ions to influence the bonding in a SiO_2 melt is mainly controlled by the ratio of the electrical charge of the ion ze to its radius r, so ze/r. If ions with rather great values of this ratio, like Sr^{2+} and Mg^{2+} and simultaneously those with the smallest value, like K^+, Cs^+, and Rb^+, are added, such a melt can separate in two melts with different degrees of polymerization on the atomistic seal. This phenomenon is called "liquation". It creates a large amount of interfaces because of the very small dimensions of the "drops" of the one liquid in the other.

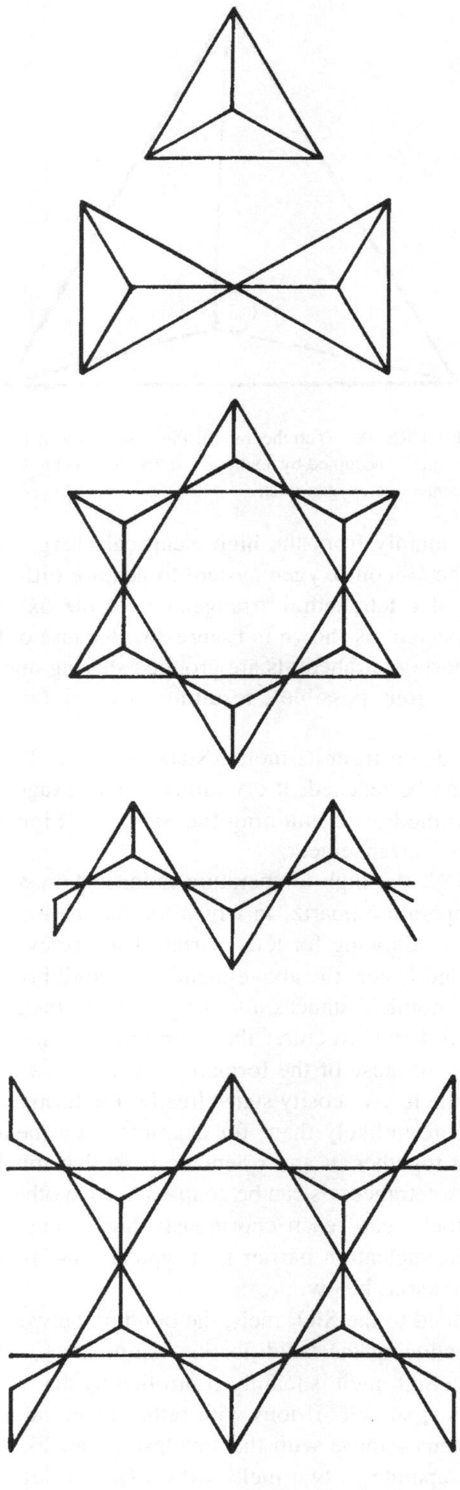

FIGURE 39. Different arrangements of tetrahedrals in silicates.

If a glass is to be devitrified it is, thus, important to create the conditions favorable for the nucleation of crystals. It is known from the observations of phase transitions and the efforts to avoid supersaturation that the introduction of inhomogeneities into a melt can assist nucleation. Such conditions exist in glasses in which liquation occurs. Glasses of this type can be found primarily in the systems Li_2O-Al_2O_3-SiO_2, Li_2O-MgO-SiO_2, and Li_2O-ZnO-SiO_2. The interfaces appearing here between the droplets of immiscible liquids or solidified immiscible liquids may be sites of a later homogeneous formation of nuclei of the crystalline phases.

Practically, this kind of devitrification is achieved by first cooling a glass of a suitable composition fast enough to obtain the usual metastable, supercooled kind of solidification. It is then reheated to a temperature beyond the range of metastability corresponding to the maximum nucleation rate of appropriate crystalline phases and, then, after the formation of nuclei, to temperatures within the range of metastability corresponding to the maximum rate of growth of already existing nuclei.

Products of this kind are called "glass ceramics". They are composed of small crystalline grains bonded by the remaining small quantities of noncrystallized glass. They can have an advantageous combination of properties, especially related to their thermal expansion behavior.

For medical applications, glasses and glass ceramics of the Na_2O-K_2O-CaO-P_2O_5-SiO_2 system have been evaluated in which crystalline phases of the tricalciumphosphate group can be expected.[4,5,21,27-29]

E. Multicomponent Ceramics

Ceramics consisting of more than one oxide have also been evaluated for medical applications during about the last 15 years. Much interest is still focused on those ceramics which closely resemble the mineral phase of bony tissue like hydroxyapatite, $Ca_5(PO_4)_3OH$; octacalciumphosphate, $Ca_4H(PO_4)_3 \times 2H_2O$; and tricalciumphosphate, $Ca_3(PO_4)$.

The CaO/P_2O_5 phase diagram is rather complicated as can be seen in Figure 40. The addition of the hydroxy ion or any other compound like fluor or chlorine increases the complexity of the system considerably.

Crystallographically the apatites belong to the hexagonal–dipyramidal group (C_{6h}) with cell dimensions in the ranges of a = 9.3 to 10.5 Å, c = 6.8 to 7.5 Å, and c/a = 0.74 to 0.71 and a density of about 3.1 to 3.2 g/cm^3. The tricalciumphosphate has the whitlockite structure (D_{3d}) with the unit cell dimensions a = 10.34 Å, c = 36.9 Å, and c/a = 3.57.

F. Ceramic Technology

While the basic manufacturing technology and methods of metals consisting of melting, casting, sometimes rolling and/or forging, and finally turning and/or cutting are familiar to nearly everybody, the manufacturing procedures of the ceramics which had been discussed above are not so widely known. However, the processes for manufacturing the oxide and similar ceramics also have an analogon in metal manufacturing: the technologies of powder metallurgy and oxide ceramics are very similar basically. In the early stages of these developments, this had even led to the introduction of the term "metal ceramic".

The main steps in this kind of technology are the starting material always is a very fine powder of the material or materials concerned. These powders are either produced by crushing, comminuting, and pulverizing solids of the materials concerned or by precipitating them from liquid solutions or sometimes also from the vapor phase. These powders are, in most cases, treated further, mixed if necessary, and prepared for shaping.

These suitably treated powders are compacted by compression in some cases to the desired shape in one step already; in other cases they are preshaped by compression only. This is mostly done with machines similar to pelletizing presses.

The shaped powder agglomerates produced by compression are transformed into a

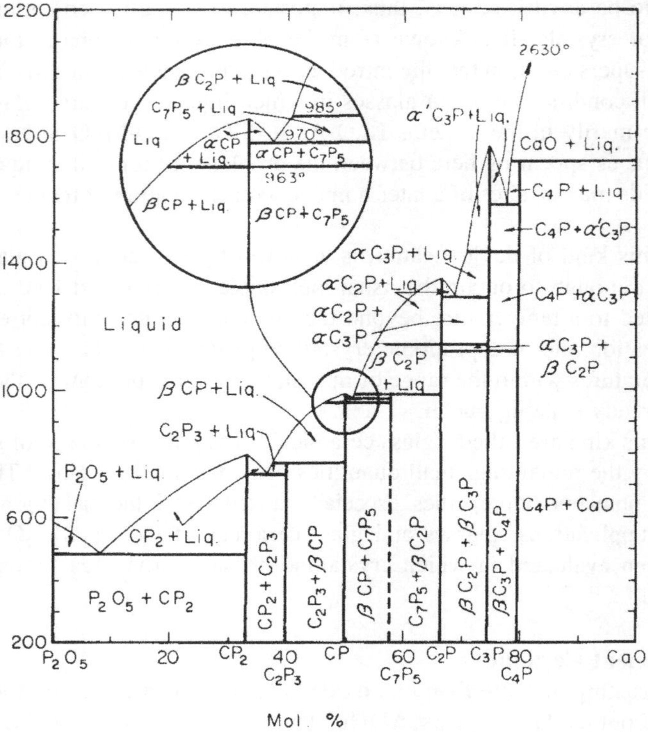

FIGURE 40. The CaO-P_2O_5 phase diagram.

polycrystalline solid by heating them to temperatures at which the atoms or ions assume a high enough mobility for enabling them to assume their equilibrium arrangement. Of course this temperature must be well below the melting temperature. This process of transformation of a powder compact into a polycrystalline body is called sintering.

This sintering is always associated with a considerable volume decrease and density increase of the body concerned, because in compressing the compacts one usually cannot reach much more than 50% of the theoretical density of the material. Thus, during sintering the volume decreases by about 50%, resulting in a linear shrinkage of about 17%. These dimensional changes limit the accuracy with which the dimensions of the sintered body can be maintained.

For further shaping and sometimes purely in order to increase the mechanical accuracy, the polycrystalline solids are further worked mechanically. With ceramics, this can be done by grinding only using diamond tools.

In all these manufacturing processes, starting from powders, the influences of the different manufacturing steps on the quality of the final product and even on the quality of intermediate products are much higher than with usual cast metal technologies. This is true the more complicated the crystalline structures and the higher their solubility for foreign atoms are. This aspect should always be kept in mind when obvious deviating tissue responses are observed with so-called identical materials. Ceramics with identical crystal structure and composition can still differ by many parameters as in grain morphology, grain size and grain size distribution, grain orientation (textures), and kind and density of lattice defects, even if the densities of the materials concerned are also identical.

G. Characteristic Properties of Ceramics

There are some essential differences between the properties of most of the metals technically used today and most of the ceramics. All of these differences are basically caused by the different kinds of bonding. As mentioned above, the metals are characterized as consisting of atoms, each of which has contributed a portion of its electrons to the "electron gas" filling the volume of the metal homogeneously with a high degree of mobility. In contrast to this situation, ceramics consist of ions with a high degree of localization of all electrical charges to the entities of the crystal lattice. There is also the necessity of maintaining the electrical neutrality within single atomic distances. This results in the basically more complicated crystal lattices of ceramics which, then, result in a much smaller mobility of lattice defects, in particular of well-organized lattice defects like edge dislocations.

The ionic structure and the much decreased availability of electrical charges result in low electrical conductivities, so that most of the ceramics discussed above are extremely good insulators, at least at room temperature. The same basic property is also the reason for the optical properties of the materials concerned. Pure corundum, for instance, does not reflect visible light at all, but has a high translucency and, as a single crystal, can be used like glasses.

The much lower heat conductivity of most of the ceramics as compared to the metals also results from the ionic structure and the lack of the easily movable electrons. The more complex crystalline structure of the ceramics together with the necessity of maintaining electrical neutrality within atomic dimensions also accounts for the different mechanical properties of the ceramics as compared to most of the metals. In discussing the plasticity of metals, it was mentioned that this is inherently related to the mobility of dislocations. The much more complicated ionic arrangements in ceramics render the mobility of dislocation in these materials much more difficult, or, in other words, demanding more energy. The threshold or activition energy for moving dislocations in ceramics is much higher. The temperatures at which all the dislocation systems, the mobility of which is needed for complete plasticity, can start moving is in the order of about 1000°C for most of the oxide ceramics.

The lack of plasticity at room temperature for most of the ceramics results in a nearly complete linear dependence of the deformation on the forces applied over the entire range up to the breakage of the material. This type of deformation behavior is called brittle. Due to the high binding forces in the ionic crystals, the deformations per unit forces are relatively small. This is expressed by a high Young's modulus or by the statement of an extremely high elasticity. In other words, these materials do maintain their shape and dimensions; they have a very high dimensional stability up to the breakage point.

While, with many metals, for instance, the tensile strength can not be used in many practical applications because of the much too large deformations at relatively low tensions already, the strength of ceramics can be fully used. Figure 41 shows the deformational behavior of metals and ceramics schematically.

For most practical applications of materials, the mechanical properties in the virgin state are of relatively small importance. What counts are the properties that are available towards the end of the required service time of the component or device concerned. Thus, the fatigue behavior must be discussed. Here, again, basic differences exist between metals and ceramics: for metals, it had been mentioned previously that the main and basic fatigue process is the consumption of the dislocations either by mutual blockage of dislocation systems or, simply, their being driven out of the crystals. In ceramics, as just mentioned above, dislocations are nearly not moving at all at room temperatures. So, the "consumption" of dislocations cannot be the main fatigue process in ceramics.

Unfortunately, there are other processes causing fatigue and, to some extent, even resulting in basically similar functional relations between strength of the materials and the time under the influence of loads or load changes. The reasons for most of the fatigue processes in ceramics are microcracks, the existence of which can hardly be avoided in any polycrystalline ceramic material.

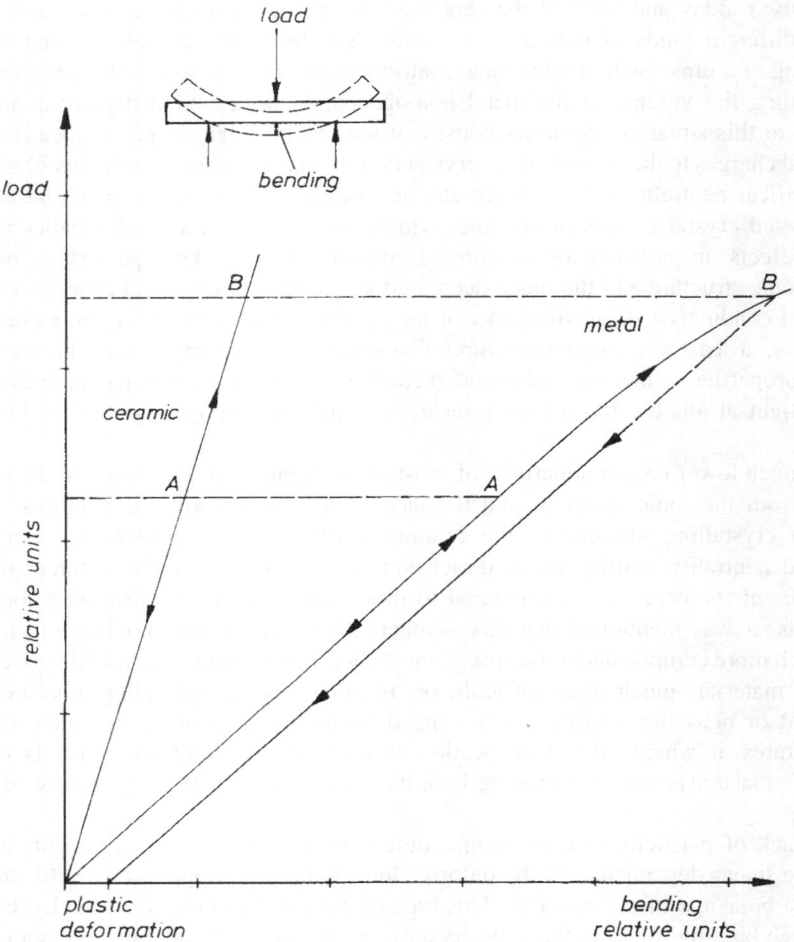

FIGURE 41. Comparison of deformation behavior of an oxide ceramic and a metal. If samples of equal cross sections are taken, a metal (like titanium) will be deformed more than twice as much under the load A. The difference of deformation will become even larger if the metal starts to flow plastically as is indicated with the load B.

These cracks are inevitably caused by the following process: the sintering temperature of these materials is, as mentioned above, in a range of high mobility of the ions concerned. This implies, of course, that the materials are highly plastic at that temperature. On subsequent cooling down, the material must pass a temperature at which the mobility of dislocations ceases. On further cooling, each tiny single crystal inside the polycrystalline body is contracting further thermally. In most of the crystalline structures of ceramics, the thermal expansion is anisotropic, being different along different axes of the lattice. As two neighboring crystals have different orientations, generally the two adjacent surfaces will try to contract slightly differently in spite of their adhering to one another. This creates a sharing force along this interface which can be calculated by multiplying the difference of the thermal expansion coefficient with the temperature difference and Young's modulus. The differences of the product of Young's modulus and the thermal expansion for the different directions of a single crystal can be plotted from one point. The end points of these "arrows" give a geometrical body which is characteristic for each crystalline material. Figure 42 shows such a body for alumina. If the tensions thus created along grain boundaries are exceeding the

FIGURE 42. Thermal expansion times Young's modulus for each lattice diection plotted from one point parallel to these directions allows construction of a body which is given here for corundum (α-Al_2O_3). (Courtesy of C. O. Leiber)

share strength, a crack will be created. The maximum size of cracks created by this process, obviously, is in the order of magnitude of the grain size.

The process described above is the main reason for the grain size dependence of the strength of ceramics and other brittle materials, which can be expressed in the following equation:

$$\sigma = \frac{A}{d^n}$$

in which σ is the strength (compressional, tensile, bend, or impact strength), A and n are constants (which maintain the temperature dependence), and d is the grain size.

The fatigue behavior of these materials is based on the following basic ideas about the influences of cracks on the strength of materials: assume a test piece with a constant cross section as indicated in Figure 43. If this test piece is put under tensional forces acting equally distributed on its right and left side, the forces inside the piece can be described by the lines of force. Because of the homogeneity, these lines have equal distances; their density over the cross section is proportional to the tension. If the piece concerned carries a crack as shown in Figure 44, the originally homogeneous arrangement of the lines of force will be severely disturbed. The cross section on which the forces are acting in the plain of the crack is much decreased. However, in addition, the distribution of the lines of force is no longer homogeneous. There is a severely increased densification of them just in front of the tip of the crack, the degree of densification depending on its radius of curvature. Thus, the tensions, expressed by the density of the lines of force in front of the tip of the crack, are increased much beyond the purely geometrically caused decrease of the load-bearing cross section.

In metals, with their ability of plastic deformation under high enough loads, the tensions

FIGURE 43. Test piece with homogeneous load line distribution inside.

FIGURE 44. Test piece with a flaw resulting in an inhomogeneous load line distribution with high concentration in front of the tip of the crack.

around the tip of a crack can lead to plastic deformations decreasing the radius of curvature and, thus, decreasing the tensions. It is this process which mainly accounts for the strength and fatigue properties of metals and their sufficiently high mechanical load-bearing ability in spite of their relatively low intrinsic binding forces. Ceramics, on the other hand, do have considerably higher binding energies as had already been mentioned above. These high strengths can actually be made available if dislocations, flaws, and cracks can be avoided as is the case in perfect single crystals or if the dimensions are small enough like in whiskers and other thin fibers. In polycrystalline materials, the above-mentioned flaws result in a considerable strength reduction.

During the last decade, however, theories have been developed allowing for a more and more detailed judgment about the strength controlling processes of such materials and even permit lifetime predictions with considerable reliability. These theories are based on particular assumptions about the processes around the tip of a crack and the predictions of these theories can be verified experimentally.[39,40]

Because the load distribution around the tip of a flaw can only be described exactly using multidimensional mathematics, the theories concerned cannot be given in detail here. However, some essential features can be described, the knowledge of which can facilitate the mutual understanding between the mechanical engineers, solid-state specialists, and biologically educated people.

The theories concerned lead to the formulation of a factor K_I describing the stress intensification around the tip of a crack. If a crack can be made to extend, obviously, the highest stress at its tip must have reached the strength of the material concerned.

$$K_I = \sigma_I \sqrt{a}\, y$$

with σ_I = strength of the material, a = crack length, and y = geometrical correction function.

To further approach these ideas, a theoretical basic experiment can be considered. First a piece of material containing a preformed slot or crack is loaded perpendicularly to the plain of the crack. If the load is increased slowly, the crack will start growing above one particular load. If, then, the load is adjusted accordingly, the crack can be kept growing slowly. Thus, a relation can be established between the crack-growing velocity, v, and the load and, from it, the stress intensity factor

$$v = A\, K_I^n$$

in which A and n are the so-called subcritical crack extension parameters.

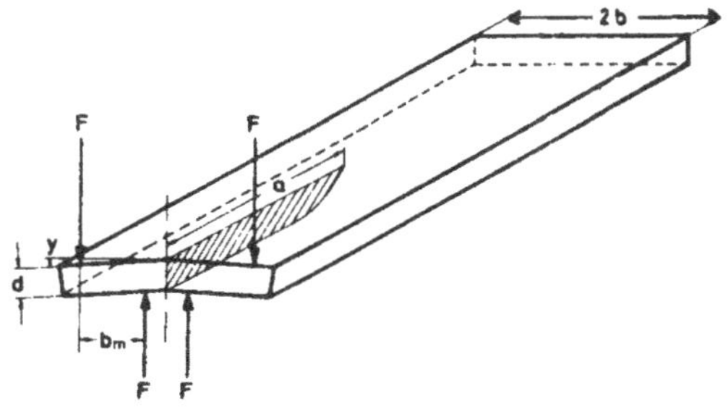

FIGURE 45. The double torsion experimental configuration for measuring the crack velocity.

This slow or so-called subcritical crack growth will continue until some particular situation arises which is given if the mechanical energy stored by the stresses around the tip of the crack is sufficient to keep the crack growing without the further contribution of outside energy. This change from subcritical crack growth into a continuous crack growth is called the critical crack extension, for which the following equation holds:

$$K_{IC} = \sigma_I \sqrt{a_C}\, y$$

with K_{IC} and a_C the critical stress intensity factor and the critical crack length, respectively.

From the above-mentioned definitions and equations, a lifetime prediction can be made using the following equation:

$$t = \frac{2\,\sigma_a^{-n}}{(n - 2)A\,Ny^2}\left(\frac{\sigma_I}{K_{IC}}\right)^{n-2}$$

with σ_a being the applied stress. This equation gives the time until which the applied stress σ_a will not lead to a critical crack extension.

Experimentally, the different values needed for evaluating the lifetime of a component can basically be derived in two ways. The first involves directly observing the propagation of a well-defined macroscopic crack in a so-called double torsion experiment, shown schematically in Figure 45. From such experiments, curves of the crack velocity vs. K_I can be derived from which, further on, the lifetime as a function of applied stresses can be predicted as shown in Figure 46.[25] The other more indirect method consists of measuring the extension of undefined microcracks in compact bending specimens by dynamic fatigue.

The validity and reliability of the theories have been tested by applying both these methods to the same material and comparing the results with each other and with results from more conventional fatigue testing procedures (measuring the Wöhler curves). The results of all these tests have shown that they have reached a considerable degree of reliability already.

Because all these tests can be performed under different surrounding media, reliable predictions of the life expectations of devices made of ceramics in biological environments can be given. So, it has already been shown that performing these tests with the specimens immersed under water results in the same degree of fatigue as in biological fluids.

V. CONCLUSIONS

In describing the main characteristics of metals and ceramics from the point of view of their structural differences, it was shown that most of them can be traced back to the differences of the types of bonds prevailing in the two types of materials: the nondirectional

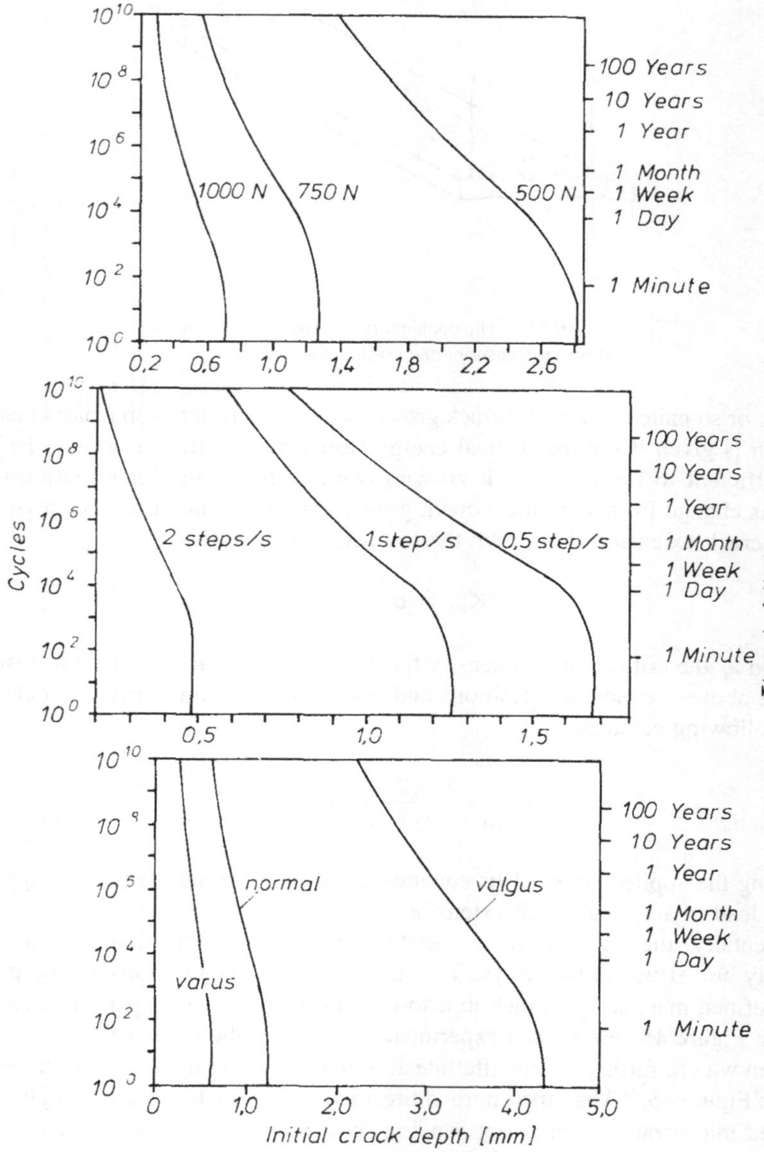

FIGURE 46. Time to failure predictions based on fracture mechanics as a function of the initial flaw size for a femoral component of a hip endoprosthesis made completely of a high-strength Al_2O_3 ceramic, assuming the initially stressed area to be in the lateral side of the neck, a double-peak walking loading, and a peak force of 1000 N. (A) For different body weights; (B) for different walking speeds; (C) for valgus, neutral, or varus position of the prosthesis in the femur. (Thanks are due to Drs. Soltész, Richter, and Seidelmann, Freiburg/Br., for making these graphs available.)

metallic bond with an abundance of freely moving electrons accounting for the relative ease with which lattice defects (dislocations) can be moved and for the electrical homogeneity of a metals surface (equal potential along the whole surface) on the one hand and for the high degree of rigidity and the more strongly absorbing surfaces of the ceramics on the other hand. The high binding energies of the least noble metals to oxygen contribute also to the differences in the corrosion behavior of the two types of materials. It is this kind of high energy oxygen bond that also accounts for the relative corrosion resistance of the less soluble metals like the chromium oxide layers on the surface of the stainless steels and the Co-based alloys and the titanium oxides on titanium and its alloys.

The close adherence of bony tissue to the hydroxyapatites and to some of the calcium phosphate ceramics could have been interpreted as just another improved version of the good "wetting" properties of ionic crystals to the components of body fluids. However, the degradability of some of these materials seemed to contradict the above-mentioned stability of oxide systems. Understanding of these phenomena has improved in recent years by the group working in Amsterdam. They have been summarized by de Groot.[19]

All the ceramics based on calcium salts with a Ca/P ratio of 1 to 2 can take hydroxyl groups into their lattices in a biological environment, thus rendering their surfaces into hydroxyapatites. However, this same process leads to a severe mechanical weakening of the affected portions of the solids concerned. If they are of small cross section like the "necks" between loosely sintered tiny particles, those particles can be disintegrated and become available for phagocytosis. Thus, the biodegradability and the ability to form a close bond to bony tissue could be traced back to the same property of the calcium phosphate ceramics, their solubility for hydroxyl ions.

The friction and wear properties of solids also correlate closely to the kind and strength of the binding forces: hardness is defined as the resistance to surface damages like scratches or indentations. Of course, this resistance strongly depends on the strength of the bonds, rendering the materials with the highest binding energies into the hardest ones: diamond and corundum next to it. Thus, the hardest ceramics, Al_2O_3, have been considered for the articulating surfaces of heavy load-bearing joints at early stages of their evaluation for medical applications already.[7,14,24] In the meantime, these early experimental findings have been confirmed clinically in all those cases in which impact loadings or, eventually temporary, edge contacts of ceramic components against other ceramic components could be avoided.[8,17,26] If materials of different hardness are articulating like in the stainless steel/polyethylene, the Co-Cr-Mo alloy/polyethylene, or the ceramic/polyethylene combination, the destruction and consumption of the softer component is delayed only by two processes: first, the presence and functioning of lubricating layers or films, the adherence of which onto ceramic surfaces again is favored by the ionic structure and the strong adherence of serum components resulting from it, and second, the lubricating action of the abraded particles of the softer component themselves as has been mentioned in discussing the graphites above.

The influences of the basic structural differences of the inorganic implant materials will be recognized in the following contributions of this volume again and again, even if not particularly mentioned. Their recognition can also contribute to the selection of other and new potential materials for medical applications and their evaluations.

REFERENCES

1. **Albrektsson, T., Branemark, P.-I., Hansson, H.-A., Ivansson, B., and Jönsson, U.,** Ultrastructural analysis of the interface zone of titanium and gold implants, in *Advances in Biomaterials*, Vol. 4, Lee, A. Y. C., Albrektsson, T., and Branemank, P.-I., Eds., John Wiley & Sons, New York, 1982, 167.
2. **Bender, B., Williams, D. B., and Notis, M. R.,** Investigation of grain-boundary segregation in ceramic oxides by analytical scanning transmission electron microscopy, *J. Am. Ceram. Soc.*, 63, 542, 1980.
3. **Benson, J.,** Presurvey on Biomedical Application of Carbons, Rocketdyne Report, R-7855, North American Rockwell, El Segundo, Ca., May 9, 1969.
4. **Blencke, B.-A., Brömer, H., and Pfeil, E.,** Rasterelektronenmikroskopische Untersuchungen der Reaktion des Knochens auf glaskeramische Implantate, *Z. Orthop. Ihre Grenzgeb.*, 112, 978, 1974.

5. **Blencke, B.-A.,** Derzeitiger Stand der klinischen Anwendung von Keramiken für den Knochen- und Gelenkersatz, *Orthop. Lect.*, 7, 43, 1978.
6. **Bokros, J. C., Gott, V. L., La Grange, L. D., Fudali, A. M., and Ramos, M. D.,** Heparin sorptivity and blood compatibility of carbon surfaces, *J. Biomed. Mater. Res.*, 4, 145, 1970.
7. **Boutin, P.,** Arthroplastie totale de la hanche par prothèse en alumine frittée, *Rev. Chir. Orthoped. Reparatrice L'Appareil Moteur (Paris)*, 58 (3), 229, 1972.
8. **Boutin, P. and Blanquaert, D.,** Le frottement alumine-alumine en chirurgie de la hanche, 1205 arthroplasties totales: avril 1970—juin 1980, *Rev. Chir. Orthoped. Reparatrice L'Appareil Moteur (Paris)*, 67, 279, 1981.
9. **Branemark, P.-I., Adell, R., Albrektsson, T., Carlsson, G., Haraldson, T., Lekholm, U., Lindqvist, L., Lindström, J., Lundqvist, S., and Rockler, B.,** Osseointegrated titanium implants in the rehabilitation of the edentulous patient, in *Advances in Biomaterials*, Vol. 4, Lee, A. Y. C., Albrektsson, T., and Branemunk, P.-I., Eds., John Wiley & Sons, New York, 1982.
10. **Brückmann, H. and Hüttinger, K. J.,** Carbon, a promising material in endoprosthetics, *Biomaterials*, 1, 67, 1980.
11. **Brückmann, H., Mäurer, H.-J., Hüttinger, K. J., Rettig, H., and Weber, U.,** New carbon materials for joint prostheses, in *Biomaterials 1980*, Winter, G. D., Gibbons, D. F., and Plenck, H., Jr., Eds., John Wiley & Sons, New York, 1982, 27.
12. **Driskell, T. D., Hassler, C. R., Tennery, V. J., McCoy, I. R., and Clarke, W. J.,** Calcium phosphate resorbable ceramics: a potential alternative to bone grafting, *J. Dent. Res.*, 52, 123, 1973.
13. **Ebert, R. and Schaldach, M.,** The applicability of rutile ceramics for cardiovascular devices, presented at 12th Int. Conf. Medical and Biol. Eng., Jerusalem, Israel, 1979.
14. **Griss, P., Krempien, B., von Andrian-Werburg, H., Heimke, G., and Fleiner, R.,** Experimentelle Untersuchung zur Gewebsverträglichkeit oxidkeramischer (Al_2O_3) Abriebteilchen, *Arch. Orthop. Unfall Chir.*, 76, 270, 1973.
15. **Griss, P., Werner, E., Heimke, G., and Raute-Kreinsen, U.,** Vergleichende experimentelle Untersuchungen von Al_2O_3-Keramik und mit mod. Bioglas (L.L. Hench) beschichtete Al_2O_3-Keramik. II. Ergebnisse mit belasteten Implantaten, *Arch. Orthop. Traumat. Surg.*, 92, 199, 1978.
16. **Griss, P., Werner, E., and Heimke, G.,** Alumina ceramic, bioglass, and silicon nitride: a comparative biocompatibility study, *Mechanical Properties of Biomaterials*, Hastings, G. W. and Williams, D. F., Eds., John Wiley & Sons, New York, 1980, 217.
17. **Griss, P. and Heimke, G.,** Five years experience with ceramic-metal-composite hip endoprostheses. I. Clinical evaluation, *Arch. Orthop. Traumat. Surg.*, 98, 157, 1981.
18. **de Groot, K.,** Bioceramics consisting of calcium phosphate salts, *Biomaterials*, 1, 47, 1980.
19. **de Groot, K.,** Degradable ceramics, in *Biocompatibility of Clinical Implant Materials*, Williams, D. F., Ed., Vol. 1, CRC Press, Boca Raton, Fla., 1981, 199.
20. **Gross, U., Brandes, J., Strunz, V., Bab, I., and Sela, J.,** The ultrastructure of the interface between glass ceramic and bone, *J. Biomed. Mater. Res.*, 15, 291, 1981.
21. **Gross, U. M., Pickartz, H., Taube, K.-G., and Strunz, V.,** Tissue reactions during the incorporation of glass ceramic material, in *Evaluation of Biomaterials*, Winter, G. D., Leray, J. L., de Groot, K., Eds., John Wiley & Sons, New York, 1980, 413.
22. **Hansen, M.,** *Constitution of Binary Alloys*, McGraw-Hill, New York, 1958.
23. **Haubold, A. D., Shim, H. S., and Bokros, J. C.,** Carbon in medical devices, in *Biocompatibility of Clinical Implant Materials*, Vol. 2, Williams, D. F., Ed., CRC Press, Boca Raton, Fla., 1981, 3.
24. **Heimke, G., Beisler, W., von Andrian-Werburg, H., Griss, P., and Krempien, B.,** Untersuchungen an Implantaten aus Al_2O_3-Keramik, *Ber. Dtsch. Keram. Ges.*, 50, 4, 1973.
25. **Heimke, G., Griss, P., Jentschura, G., and Werner, E.,** Mechanical testing of ceramic materials and components for bone and joint replacements, in *Mechanical Properties of Biomaterials*, Hastings, G. W. and Williams, D. F., Eds., John Wiley & Sons, New York, 1980, 129.
26. **Heimke, G. and Griss, P.,** Five years experience with ceramic-metal-composite hip endoprosthese. II. Mechanical evaluations and improvements, *Arch. Orthop. Traumat. Surg.*, 98, 165, 1981.
27. **Hench, L. L., Splinter, R. J., Allen, W. C., and Greenlee, T. K.,** Bonding mechanisms at the interface of ceramic prosthetic material, *J. Biomed. Mater. Res. Symp.*, 2, 117, 1971.
28. **Hench, L. L.,** Development of a new biomaterial prosthetic device, in *Orthopedic Mechanics, Procedures and Devices*, Ghista, D. H. and Roaf, R., Eds., Academic Press, New York, 1978, 287.
29. **Hench, L. L., Petty, R. W., and Piotrowski, G.,** An Investigation of Bonding Mechanisms at the Interface of a Prosthetic Material, Rep. 9, U.S. Army R & D Command, Contract No. DADA 17-70-C-0001, Washington, D.C., 1978.
30. **Hüttinger, K. J. and Mäurer, H.-J.,** Tribiological properties of carbon materials in artificial joints, *Biomaterials 1980*, Winter, G. D., Gibbons, D. F., and Plenck, H., Jr., Eds., John Wiley & Sons, New York, 1982, 177.

31. **Jarcho, M.,** Calcium phosphate as hard tissue prosthetics, *Clin. Orthop.*, 157, 259, 1981.
32. **Krüger, J.,** Fundamental aspects of the corrosion of metallic implants, in *Corrosion and Degradation of Implant Materials*, Syrett, B. C. and Acharga, A., Eds., American Society for Testing and Materials, Philadelphia, 1979, 107.
33. **Monticelli, Pr., Santori, Pr., Sandrolini Costese, S., and Sandrolini Costesi, M.,** First considerations on the use of ZrO_2 in hip prosthesis stem coating, Paper VIII, presented at Annu. Meeting Eur. Soc. Artificial Organs, Copenhagen, August 1981.
34. **Mountvala, A. J.,** Surface properties of bioceramic vascular prostheses, *Ceram. Bull.*, 52, 479, 1973.
35. **Neugebauer, R., Burri, C., Claes, L., and Helbing, G.,** Trap-door fixation of braided carbon fiber knee joint ligament prostheses, in *Biomaterials 1980*, Winter, G. D., Gibbons, D. F., and Plenck, H., Jr., Eds., John Wiley & Sons, New York, 1982, 287.
36. **Österholm, H. H. and Day, D. E.,** Calcium migration in dense alumina aged in water and physiological media, *Ceram. Bull.*, 60, 955, 1981a.
37. **Österhom, H. H. and Day, D. E.,** Dense alumina aged in vivo, *J. Biomed. Mater. Res.*, 15, 279, 1981b.
38. **Osborn, J. F., Newesely, H., Werhahn, C., Brückmann, H., and Gabriel, E.,** Doped ceramics — a new approach to the dynamics of osteogenesis following calcium phosphate ceramic implantation, in *Advances in Biomaterials*, Vol. 4, Lee, A. Y. C., Albrektsson, T., and Branemark, P.-I., Eds., John Wiley & Sons, New York, 1982, 219.
39. **Pabst, R., Wichmann, G., and Elssner, G.,** Fracture toughness of alumina implant materials as a function of environment, in *Evaluation of Biomaterials*, Winter, G. D., Leray, J. L., and de Groot, K., Eds., John Wiley & Sons, New York, 1980, 219.
40. **Richter, H., Seidelmann, U., and Soltész, U.,** Slow crack growth and failure prediction for alumina in simulated physiological media, in *Evaluation of Biomaterials*, Winter, G. D., Leray, J. L., and de Groot, K., Eds., John Wiley & Sons, New York, 1980, 227.
41. **Wang, H. A. and Kröger, F. A.,** Chemical diffusion in polycrystalline Al_2O_3, *J. Am. Ceram. Soc.*, 63, 613, 1980.
42. **Williams, D. F.,** Corrosion of implant materials, *Am. Rev. Mater. Sci.*, 6, 237, 1976.
43. **Williams, D. F.,** Titanium and titanium alloys, in *Biocompatibility of Clinical Implant Materials*, Vol. 1, Williams, D. F., Ed., CRC Press, Boca Raton, Fla., 1981, 9.
44. **Zwicker, U., Bühler, K., Müller, R., Beck, H., Schmid, H. J., and Ferstl, J.,** Mechanical properties and tissue reaction of a titanium alloy for implant material, paper presented at 4th Int. Conf. on Titanium Kyoto, Japan, May 19 to 21, 1980.
45. **Müller, E.,** personal communication.
46. **Brandner, H.,** personal communication.

Chapter 3

STRUCTURAL CHARACTERISTICS OF METALS AND CERAMICS

O. Arkens and P. Ducheyne

TABLE OF CONTENTS

I. INTRODUCTION

The information that is obtained from a microstructural study of a metallic or ceramic biomaterial can be listed under two headings: first, there are the data on the morphology and topography of the surface, the crystal structure, and the distribution of the present phases, and second, there are the results relating the chemical composition to the observed microstructural features. The commercially available techniques used for such a microstructural study are dominated by first, microscopy and second, spectroscopy.

Microscopy can be defined as the interpretative application of microscope magnification to the study of materials that cannot be properly seen by the unaided eye. Spectroscopy can be defined as the branch of physics concerned with the production, the measurement, and the interpretation of energy spectra arising from either emission or absorption of radiant energy or particles from a substance upon bombardment by electromagnetic radiation, electrons, neutrons, protons, ions, or neutrals. Spectroscopical techniques are used to investigate atomic, molecular, or solid-phase structures and to analyze the chemical composition.

It is the aim of this chapter to discuss the more advanced methods in microscopy and spectroscopy related to the study of phenomena in the micron range, microtopography, bulk microanalysis, and analysis of thin surface layers. The relevance to biomaterials studies will be indicated.

Several direct methods have been developed over the past decade to investigate the surface structure of metallic and ceramic materials. In the field of microscopy the scanning electron microscope (SEM) has captured the imagination of people in all areas of scientific endeavor. The use of spectrometers on the SEM transforms this instrument in an electron probe microanalyzer (EPMA), historically a predecessor of the SEM, but using advanced spectrometers and microprocessors, an outstanding instrument for qualitative and quantitative chemical analysis in the micron range.

For the investigation of biocompatibility, the knowledge of the composition of the real surface layers (a few monolayers) of the materials used is important. Spectroscopical techniques such as the Auger electron spectroscopy (AES), photoelectron spectroscopy for chemical analysis (ESCA), and secondary ion mass spectrometry (SIMS) are important analytical tools.

II. MICROSCOPY

A. The Scanning Electron Microscope and the Electron Probe Microanalyzer

The large number of articles appearing in many of the scientific journals describing the technique of the scanning electron microscope attests to the very great interest this exciting instrument has generated. One would probably have to go back to the days when the transmission electron microscope (TEM) first arrived on the scene to find a comparable response. It is not hard to identify the reasons for this high level of interest, nor is it difficult to explain why the SEM seems to overshadow the electron microprobe even though these two instruments can properly be considered as but two aspects of a single analytical tool. To put it simply, because "one picture is worth a thousand words" (Figure 1), the SEM can provide pictures of practically every kind of solid surface and with a continuous range of magnification from as small as 5 × magnification up to 100,000 ×.

Furthermore, practically all kinds of natural surfaces can be observed, including those of live insects; and should the surface be one that does not adequately conduct away the impinging electrical current (i.e., most minerals, ceramics, and biological materials), then a thin coating of conductive materials such as gold, aluminum, or carbon is sufficient to make them so. Now, although natural surfaces are most often very rough, the SEM processes the property of a large depth of field which permits precise focusing over much larger depths

FIGURE 1. Scanning electron micrographs. (a) A fracture surface of a femoral component of a total hip prosthesis failed in fatigue[1] (magnification × 20); (b) a porous metal fiber coating, allowing osseous tissue ingrowth[2] (magnification × 50); (c) an individual fiber of a porous metal coating mechanically tested until failure; this micrograph illustrates the ductile behavior of the porous metal coating[3] (magnification × 2000).

than conventional optical microscopy. Figure 1a to c shows some typical examples of SEM studies on biomaterials.

What kind of information does microscopy provide us? Classically, it provides a representation of the surface, within the limits of the depth of field (the topography). This may reveal some idea of the constituents. If the material is multiphase, perhaps the geometrical relationship between the phases can be inferred. Much more information such as the composition of the phases, their crystal structures, their relative crystallographic orientation, the structure of the boundaries, and the presence and nature of defects in the material is frequently required. It is always desired to see these things at larger and larger magnification. The list of physical properties need not stop here. One can add to it information relative to electric and magnetic properties, chemical behavior, etc. However, the starting point is obviously composition and structure.

The distinguishing feature of the scanning microscope and the microprobe is that the images are formed sequentially i.e., point-by-point. This contrasts with optical and transmission electron microscopy where the light rays or the electrons arrive, as illustrated in Figure 2, at the imaging surface and simultaneously over the entire surface. Beam scanning images are most frequently described in terms of television pictures because here, too, an image is formed by means of a moving beam which scans the object in a raster pattern.

In the SEM and EPMA two beams are involved, an electron beam in the main column of the instrument and another in the cathode ray tube (CRT) on whose face the image is developed. As the first beam moves across the object or sample, the interaction of the high energy electrons on the solid results in several effects whose change from point to point is measured by a suitable sensing device which translates that effect into an electrical signal. The signal can then be processed: amplified, recorded, integrated, differentiated, or even transmitted to a computer if so desired. An image is created when the formation signal is amplified and then used to control the intensity of second beam in the CRT. Because the

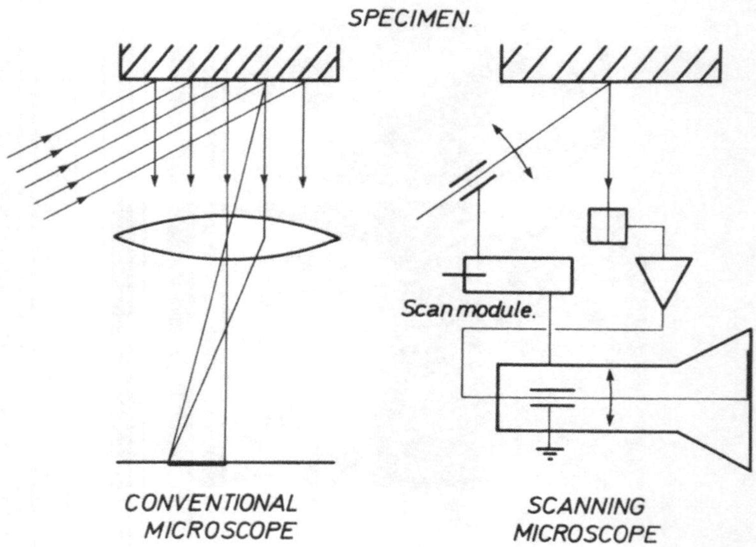

FIGURE 2. Schematic comparison of conventional and scanning microscope.

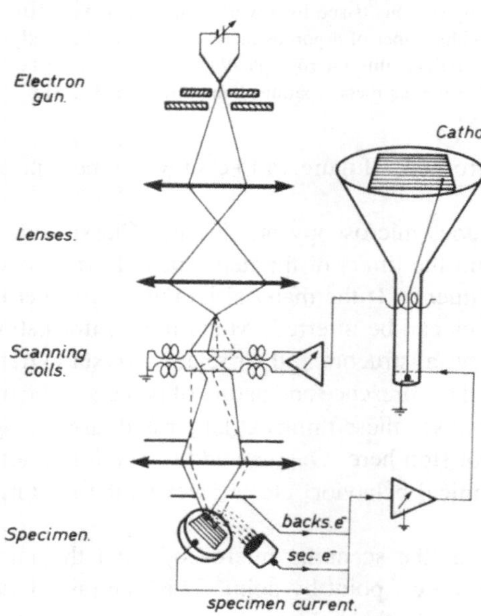

FIGURE 3. Schematic representation of SEM.

two beams are synchronized, a point on the surface of the CRT corresponds to a point on the surface of the object. Magnification is now a very simple concept: the ratio of the size of the area on the display tube to that scanned on the object.

The formation of the electron beam is shown in Figure 3. Two or three electromagnetic lenses are used to demagnify the source of electrons which come from the electron gun to a fine beam up to approximately 3 nm in diameter. The gun is also a sort of electrostatic lens. Electrons are emitted from the tip of a fine tungsten wire bent into a hair pin shape and accelerated by the power supply to the high voltage required. Depending on the sample and the kind of information required, the high voltage used may vary from 1 to 50 kV. The

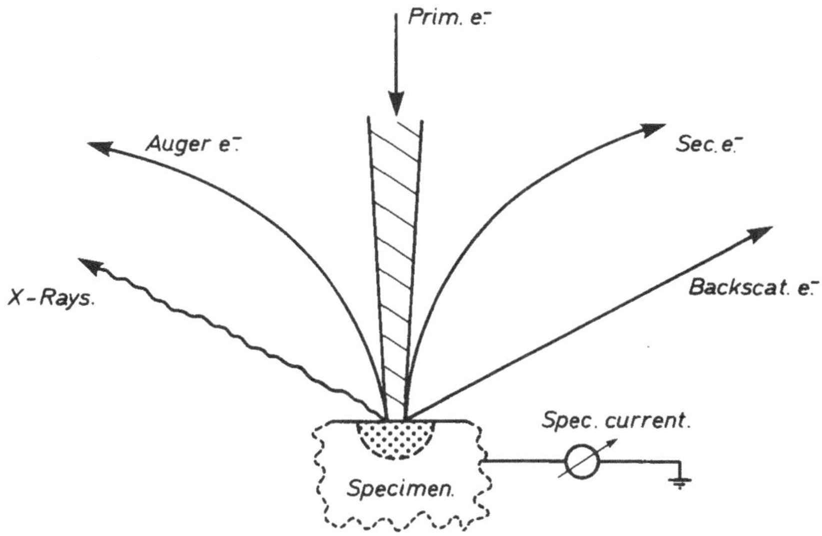

FIGURE 4. Summary of the emitted signals.

condensor lenses serve to control the magnitude of the beam current which can vary from 10^{-5} to 10^{-12} A as well as to demagnify the source. The final (objective) lens is the most critical one, since it determines the minimum beam diameter on the specimen. Incorporated in this lens are additional magnetic coils that provide the scanning motion of the beam (scanning coils) or produce small magnetic fields to counteract those already present that distort the beam out of desired circular form (stigmator coils).

Several information modes result from the high energy impact of an electron with a solid (Figure 4). Some of the incident electrons, a number that depends on the average atomic number of the sample at the point of impact, will be reflected from the surface with only slight energy loss. These back-scattered electrons (BE) can be detected by either a scintillator crystal-photomultiplier (SPM) combination or a n-type Si surface barrier counter. Both types of detectors produce an immediate and very rapid initial amplification of the electron signal. The SPM does so by converting the electron into a photon which then strikes the photosensitive surface of the photomultiplier (PM) tube, producing an avalanche of electrons. The impact of the high energy electron on the Si diode produces electron hole pairs. The rapid collection of these at the terminals results in a useable current pulse. The pulses converted to the CRT give an image that in different levels of brightness shows the different mean atomic numbers on the scanned surface (Figure 5).

In addition, an even larger number of very low energy electrons termed secondaries are emitted at the point of impact. These electrons, with energies less than 50 eV, can be drawn away from the surface of the sample by an appropriately positioned electrostatic potential. Acceleration of these electrons to 10 keV or more will permit them to be detected by the SPM detector. The number of electrons collected from a point on the surface depends strongly on the relative orientation of the surface with respect to the beam and the collector. It also depends to a lesser extent on the composition and on the crystal structure of the irradiated volume.

Another electron signal is frequently used, namely, those electrons which are absorbed corresponding to the amount of impinging electrons minus those lost due to scattering. By direct amplification, this fraction also produces an image of the surface. Amplifier limitations restrict this mode of operation to the higher current levels of the incident beam.

It is well known that the impact of an electron of sufficiently high energies with an atom

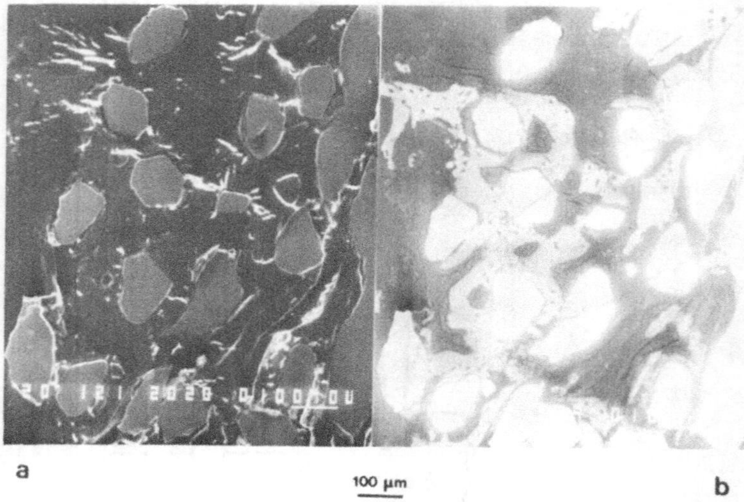

FIGURE 5. (a) Secondary electron image of metallic fibers in cross section after bone and tissue ingrowth; (b) BE image revealing the metallic fibers (white) and the surrounding bone (grey) and tissue (dark)

will generate X-rays, most of which are characteristic of the element being bombarded. The actual volume of X-rays generated will depend on the critical excitation energy of the atoms present. The volume of X-rays generated roughly covers the volume of penetration of primary electrons ($>1\ \mu m^3$). Detection and identification of X-rays in the micron range is in fact the basis of the EPMA. There are two ways by which the characteristic X-rays are identified: by their wavelength (WDS, wavelength dispersive spectrometer) or by their energy (EDS, energy dispersion spectrometer). In the first method, an X-ray crystal of a known spacing (d) separates the various wavelengths in accordance with Bragg's law for diffraction ($n\lambda = 2d \sin \theta$) before they enter the detector. So, at a certain diffraction angle θ, X-rays of a fixed wavelength (λ) are reflected to the detector. The pulses from the detector can be amplified and permit a counting of the number of the generated X-rays. In the second procedure, a detector is used which produces a pulse whose size is strictly proportional to the photon energy. In the first instance, the X-ray detector is a conventional proportional counter, whereas in the second it is a lithium drifted silicon (SiLi) semiconductor device. The crystal spectrometer has a much higher spectral resolving power, but can measure only one wavelength at a time. The solid-state detectors, as they are frequently called, have a much higher collection efficiency and can measure the entire spectrum during one single period (Figure 6).

The crystal spectrometers can be used to measure the radiation from the light elements, down to boron, whereas the presently available solid-state detectors are limited to the elements from sodium and higher up in the periodic table.

We are now in a position to relate the mode of operation with the kind of information it can provide. This is presented in the following table.

Signal used	Operative mode	Information provided
Secondary electron	Emissive	Topography, surface voltage variation
BE	Reflective	Composition variation, crystallographic orientation
Absorbed electrons	Microanalysis	Composition determination, element discrimination (Figure 7)

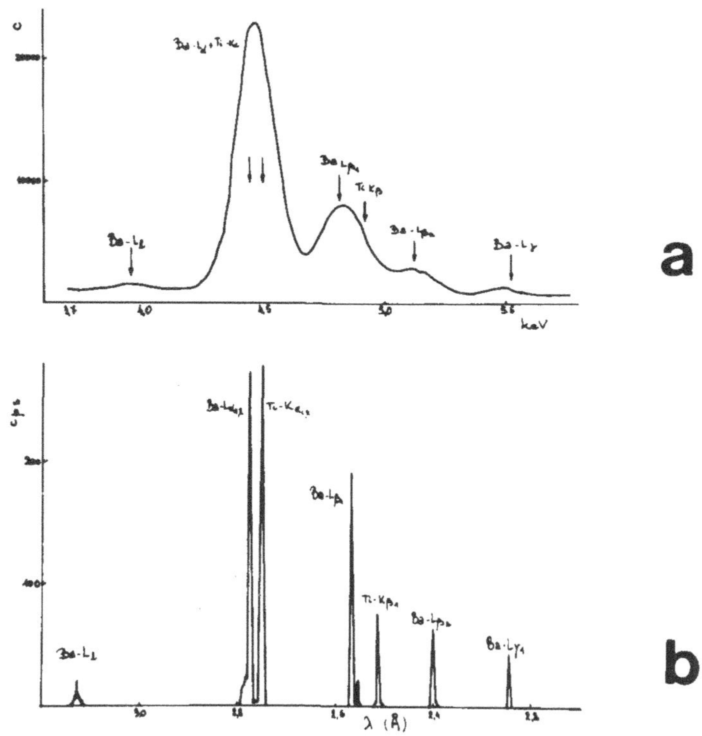

FIGURE 6. Spectra obtained by energy dispersive (ED) spectrometry (a) and wavelength dispersive (WD) spectrometry (b), showing the differences in resolution and peak/background ratio specimens: $BaTiO_3$.

B. The Scanning Electron Microscope or the Microprobe

Whereas the SEM and the EPMA have developed as two distinct instruments, the differences are more historical than instrumental. It is with respect to applications that an important difference exists. The EPMA has as its primary goal the accurate determination of elemental composition. A necessary condition to do so is a very smooth surface. Thus, most specimens for quantitative microprobe analysis will have had their interesting topographical features removed to the extent that this was possible using the various polishing procedures. If, however, the compositional information desired would be more qualitative, more irregular surfaces can still be investigated. The constructive requirements for both types of analysis — quantitative or qualitative — can be met within a single instrument, although there are some difficulties in arranging the several systems (electron optics, X-ray optics, visible light optics, and other detectors) in accordance with the optimum operation conditions of each. Thus, an instrument would be considered SEM or an EPMA depending on whether the arrangement of detectors favors the requirements of collection of secondary electrons (high resolution imaging, SEM) or optimal X-ray detection (EPMA).

Quantitative composition determination on micron-sized volumes is the special domain of the EPMA. The state of the art of composition determination has progressed considerably in recent years. It is possible to detect radiation from all elements down to and including boron. The wavelength of the B K_α line is 6.7 nm. At present, sensitivities of the order of a few tenths of a percent are possible when reference standards are used. However, there are situations such as one where the signal of a light element is heavily attenuated by the other elements present in the matrix, so that the sensitivity is strongly reduced. The situation improves as one proceeds to higher atomic numbers as both peak intensity per microampere

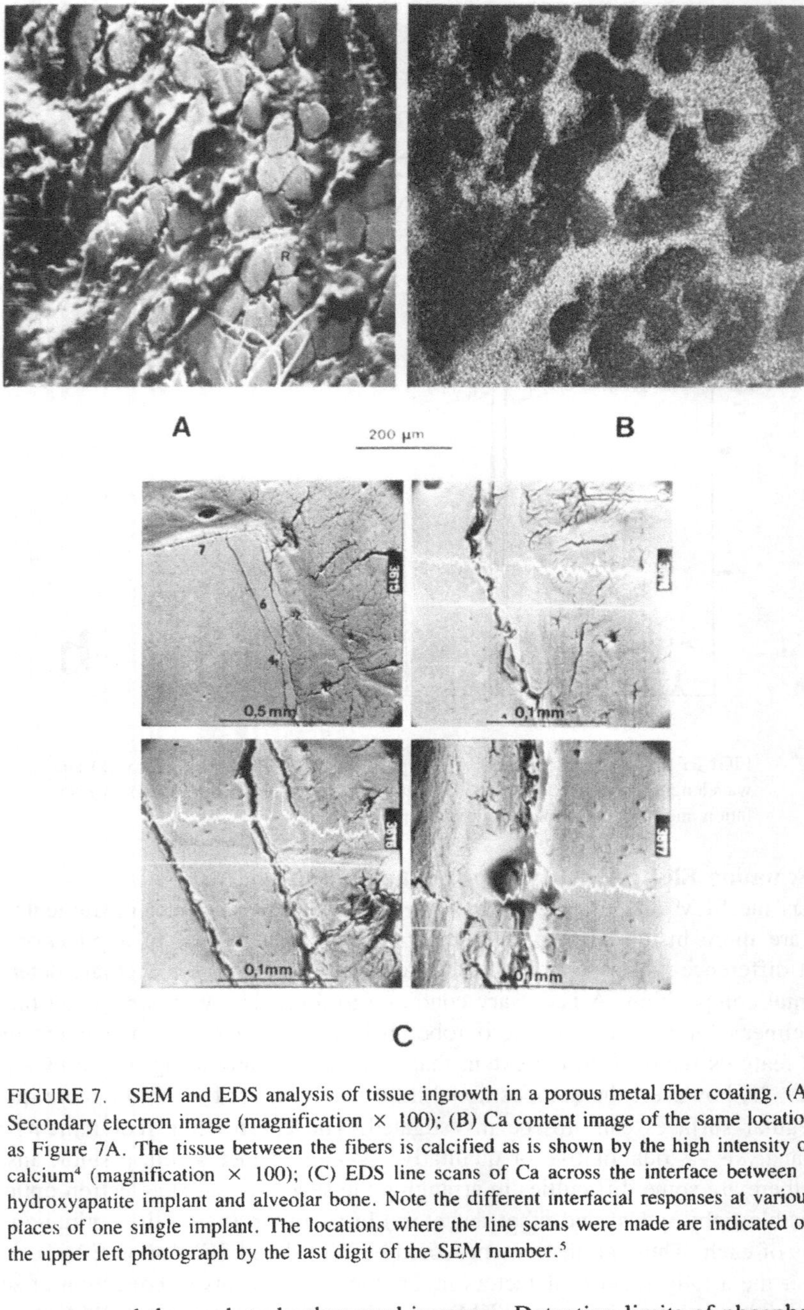

FIGURE 7. SEM and EDS analysis of tissue ingrowth in a porous metal fiber coating. (A) Secondary electron image (magnification × 100); (B) Ca content image of the same location as Figure 7A. The tissue between the fibers is calcified as is shown by the high intensity of calcium[4] (magnification × 100); (C) EDS line scans of Ca across the interface between a hydroxyapatite implant and alveolar bone. Note the different interfacial responses at various places of one single implant. The locations where the line scans were made are indicated on the upper left photograph by the last digit of the SEM number.[5]

of probe current and the peak-to-background increase. Detection limits of phosphor in steel as low as several hundredths of a percent have been reported. For detection and measurement of oxygen and nitrogen, the situation deteriorates somewhat since the K wavelength of these elements is usually strongly absorbed by the beryllium window of the flow proportional counter, which is used to detect this radiation. As new window materials are developed, there should be corresponding improvements in both signal intensity and signal-to-noise ratios.

Instrument automation and correcting procedures by microprocessors and computers have reduced the analysis time and improved the quantitative results obtained in the EPMA.

Precision measurements can be made if reliable standards of approximately the same composition are available. Then the accuracy is limited only by the usual X-ray counting statistics and the uncertainty in the standards themselves. In common use today is the ZAF correcting approach, where analytical expressions are used for calculating the atomic number (Z), absorption (A), and secondary fluorescence (F) and combinations in a multielement target. The EPMA can be used to measure trace amounts within the limitation of counting statistics as affected by peak counting rates and peak to background ratios. Under favorable conditions, trace amounts of 10 to 100 ppm can be measured.

In the biological sciences one finds the most dramatic examples for the SEM where the topographical (or emissive) mode is the major form of use of the instrument. A distinction between internal and external structure may be useful here. The internal structure can be revealed by thin sections, staining, and in addition transmission electron microscopy. Even though the dimensions of such a structure are often on the submicron scale, to perform analyses for the distribution of the elements of interest, good sample preparation is difficult and is the main reason for limited analytical results. In addition, biological material is not very dense, so that the electron beam accelerating voltage must be kept low to retain a small analyzing volume. The elements of interest are only present in small quantities and are almost all found in the low Z end of the periodic table where detection problems are greater. The nondispersion X-ray spectrometer using the drifted Si(Li) semiconductor device as a photon detector can operate effectively at probe current levels compatible with the SEM probe diameter of 10 nm. The X-ray generating volume will still be kept small (5 to 10 keV) since many of the elements of greatest interest (from Na to Ca) have critical excitation values well below this range (1.3 to 4.0 keV). The low current levels have another advantage in that sample deterioration will be less marked using such lower voltages and low currents. As mentioned, the limitations of the solid-state nondispersion detector are its relatively poor spectral resolution (as compared to the dispersive crystal spectrometer) and the inability to detect and resolve radiation below $Z = 11$ (Na). However, an important advantage that accrues with this type of detector is that the entire spectrum is obtained during one counting period which yields very easy qualitative comparisons of different locations on a specimen.

Installation of this detector on a TEM, which can be done easily, allows the analysis of thin sections; if this conventional TEM has scanning facilities, including fine beam focusing, one can obtain local analytical results in the nanometer range.

III. THE SURFACE ANALYSIS OF SOLIDS

The electron microprobe and the SEM equipped with spectrometers report analytical information from a depth of at least 1 μm. Although the sensitivity of the EPMA is adequate for detecting atoms from the upper few atomic layers, no information on location and concentration would be obtained under these instrument operation conditions. So the atoms which make up a solid can be divided in two classes: those which form the surface and those which are contained within its volume.

In domains where the surface phenomena are of primary importance, such as chemical surface reactions, the composition and the structure of the outermost atomic layers determine the characteristics of the solid. The photons, electrons, neutral particles, and ions emitted from the surface layer are all potential information carriers. Their emission can be stimulated by such means as bombarding the surface with photons, electrons, ions, or neutral particles, by application of high electrical fields, or by heat. In general, emitted photons and electrons are analyzed according to their energy, and atomic particles are analyzed according to their mass.

Presently used techniques are the emission and detection of photoelectrons, secondary electrons, and secondary ions. The conventional denominations used are ESCA, electron

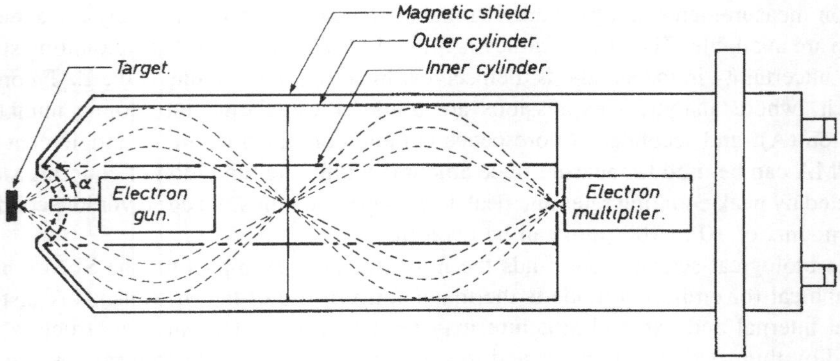

FIGURE 8. Double pass cylindrical mirror analyzer.

spectroscopy for chemical analysis; AES, Auger electron spectroscopy; and SIMS, secondary ion mass spectrometry. The procedure for the surface analysis can be summarized in the following table.

	Emission		
Stimulation	**Photons**	**Electrons**	**Ions**
Electrons	ESCA	AES	
Ions			SIMS

A. Auger Electron Spectroscopy (AES)

When a beam of electrons of sufficient energy strikes a solid surface, the so-called secondary electrons are emitted. Their sources of origin and the processes responsible for their emission are reflected in their energy distribution and can be determined by means of energy distribution analysis. Such determination can be carried out by utilizing a cylindrical analyzer (Figure 8). This apparatus allows the target to be bombarded with electrons either vertically or at grazing incidence. At a certain deflecting potential, U, the only secondary electrons to pass through the deflecting fields are those which have an energy, E, within a given range, ΔE. This energy, E, is proportional to the deflecting voltage, U. The energy interval, ΔE, is determined by the analyzer geometry and is proportional to E. By scanning the deflecting voltage, U, and hence E, an energy distribution, N(E), can be directly recorded; this, however, relates to the energy-dependent energy interval, ΔE. If the "fine structure" of the energy distribution of the emitted electrons is required to be investigated, it is convenient to pass from the function, N(E), to its derivative. This can be obtained as a direct output signal of the analyzing system (fed into a recorder) if a high-frequency AC voltage is superposed during the scan of deflecting voltage and the output of the scan is fed into an appropriately controlled lock-in amplifier (Figure 9).

These relatively sharp lines are superimposed on a broad background due to secondary electrons which are emitted over a wide energy range as a result of multiple energy losses. The detected secondary electrons are called Auger electrons.

Auger electrons result from the transition of an excited atom (ionized in a inner shell) to a lower energy state (Figure 10). A primary electron with energy, E_p, ionizes an atom in an inner level, E_1. An electron from a higher level, E_2, fills out this level. The energy, $|E_1 - E_2|$, thus released can either be emitted as a photon (characteristic X-radiation) or be transmitted to another electron in an energy level, E_3, which can then leave the solid as an Auger electron, with the characteristic energy $E_a = |E_1| - |E_2| - |E_3|$. The width of the Auger peaks depends on the resolving power of the analyzer and on the natural energy

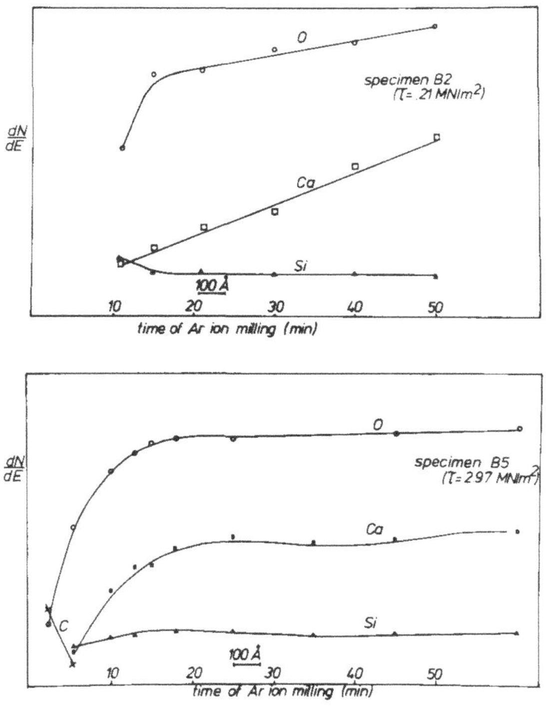

FIGURE 9. Auger electron spectroscopical measurements on Bioglass specimens with different surface preparation and subsequent different bonding behavior. Note the difference in Ca and O concentration.[6]

spread of the Auger electrons, predetermined by the natural width of the energy levels involved in the transition process.

The symbols used to designate the Auger transition process are based on those employed in X-ray spectroscopy. Therefore, a KL_1L_2 Auger transition is one where the primary ionization has taken place in the K shell and, after emission of the Auger electron, an atom ionized in the L_1 and L_2 levels remains.

The intensity of an Auger line will be determined by the intensity of the primary ionization and the probability of a subsequent Auger transition. The escape depth of Auger electrons depends on their energy and on the nature of the emitting material. The necessary condition for the Auger electrons to leave the solid with their characteristic energy E_a is that they should lose no energy (while escaping from their place of origin) apart from the work function. If other energy losses occur, these electrons will no longer have the energy corresponding to the Auger transition and hence, will not contribute to the relevant Auger peak. Hence, Auger electrons satisfy the main requirements for application in surface analysis: they originate only from the region of the outermost atomic layers and, by means of their energy, they provide direct information on their chemical composition.

The most important characteristics of AES as an analytical method can be summarized as follows:

1. The elements present in a surface region of approximately five atomic layers are detected. In general, a given element is represented by a number of lines in the spectrum, corresponding to various Auger transitions.
2. Hydrogen and helium cannot be detected because no Auger transitions occur with these elements.

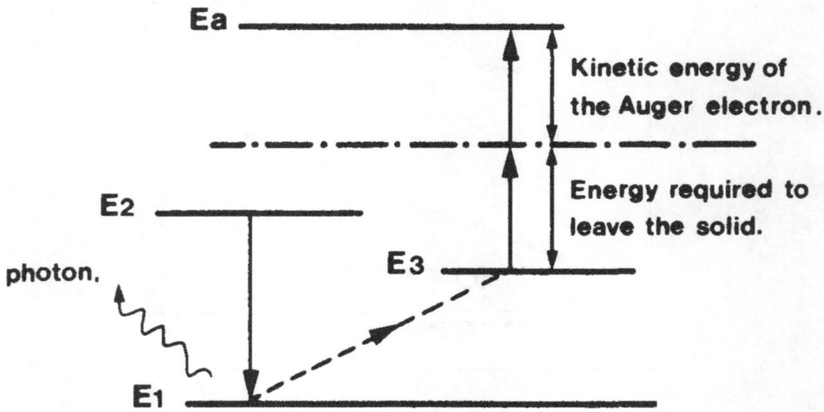

FIGURE 10. Schematic illustration of the Auger effect.

3. Under favorable conditions, the smallest traceable concentration of an element is less than 10^{-3} monolayer equivalents, which can be distributed over the determined depth range.
4. The exact position and shape of an Auger line are determined by the binding state. Unfortunately, the identification of a chemical compound on the basis of this chemical shift has been successful only in special cases.
5. The high transmission of the cylindrical analyzer enables oscillographic recording of spectra.
6. Focusing the primary electron beam on the target, and scanning this beam over the surface, allow a very local analysis and imaging of elemental distributions over the analyzed area. This is called scanning Auger spectroscopy (SAM).

B. Photoelectron Spectroscopy

The next method generally applicable for surface analysis which will be considered in greater detail is photoelectron spectroscopy, i.e., the analysis of the energy distribution of the electrons emitted from a solid upon bombardment with photons (X-rays). This technique has been developed by Siegbahn and co-workers to be an extremely efficient method for structure analysis. However, its application to surface analysis is still very much in the initial stage.

When a solid surface is bombarded with photons of sufficiently high energy, photoelectrons are emitted from inner energy levels. An atom in energy level E_1, ionized by a photon with energy $h\nu$, emits an electron with energy $h\nu - E_1$. The ionized atom can change to a lower energy state through the emission of X-radiation or through an Auger transition. In the latter case, not only the photoelectrons, but also electrons resulting from the Auger transition, will be emitted.

For a given energy of ionizing X-radiations, the energy of a photoelectron will be determined by only one energy level of the emitting atom, whereas the energy of the Auger electrons will be determined by three such levels. Therefore, if the energy of the irradiating photons is sufficiently uniform, photoelectrons will in general be emitted with considerably less energy spread than Auger electrons. This in turn means that any displacement (chemical shift) of the energy levels in the atoms due to their specific chemical state will be considerably more conspicuous in the photoelectron spectrum than in the Auger spectrum.

In order to be able to determine these chemical shifts, a photoelectron spectrometer must have better energy resolution than an Auger electron spectrometer and, in view of the low intensity of the photoelectron currents, it should also have high transmission. This is achieved

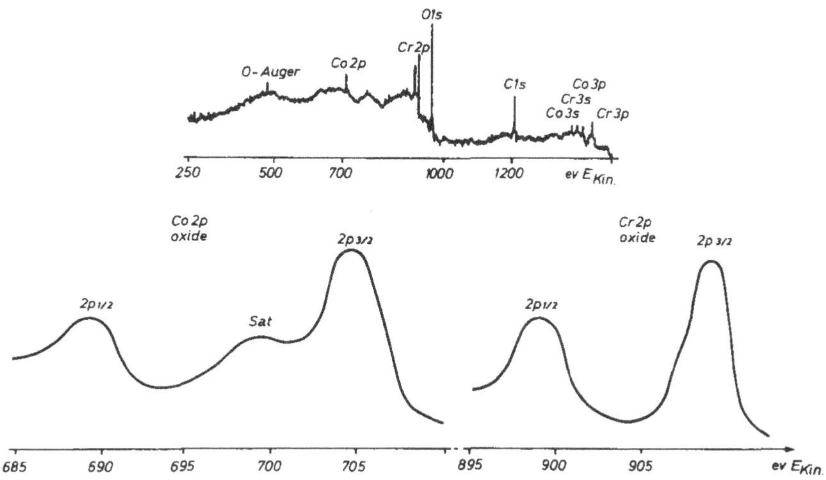

FIGURE 11. ESCA analysis of the oxide present on cast cobalt-chromium-molybdenum alloys. The oxide on a surface that is freshly polished and heated in air, 450°C for 10 min, is a mixed Co and Cr oxide.[7]

as follows: the photons produced in an X-ray source impinge on the target; the photoelectrons emitted are retarded in a lens system and enter a spherical analyzer which is set at a fixed energy. Recording of an energy distribution is possible by changing the retarding voltage in the lens system (Figure 11).

The most important characteristics of photoelectron spectroscopy as a surface analysis method can be summarized as follows:

1. In comparison with AES, the most important feature is its ability to determine, from the chemical shift, the chemical environment in which the element is present on the surface of the solid.
2. The energy loss and the free emission depth of photoelectrons is greater than that of Auger electrons, due to their generally higher energy.
3. The average information depth is of the order of ten atomic layers.
4. The limit of detection for surface atoms is approximately 1% of a monolayer.
5. Due to the lower intensity of the photoelectron currents, the time required to record a spectrum is considerably longer than in AES.
6. Hydrogen cannot be detected.

C. Secondary Ion Mass Spectroscopy

This method makes use of the fact that when a solid surface is bombarded with ions of sufficiently high energy, it will emit not only photons, electrons, and neutral particles, but also positive and negative secondary ions characteristic for the chemical composition of the emitting surface. This secondary ion emission effect has been used for a considerable time in a surface cleaning device or in an apparatus for deposition of thin layers (sputtering).

With the use of high primary ion current densities, this method has also found various applications in solid-state mass spectroscopy during the past 10 years. However, since the surface layer is removed by sputtering during an intense ion bombardment, SIMS has until recently been regarded as unsuitable for the analysis of individual monomolecular layers. The introduction of the "static method" has reduced sputtering to a negligible level by the use of extremely low primary ion current densities, so that it has become virtually a non-destructive method for the analysis of individual monomolecular layers.

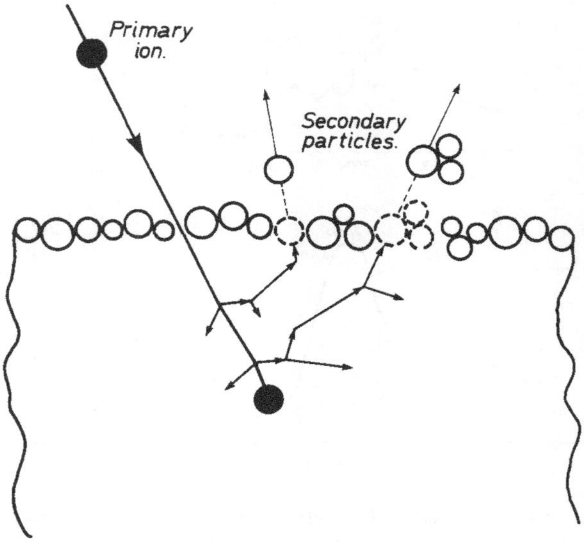

FIGURE 12. Ion-matter interactions in SIMS.

The emission of a secondary ion is the result of a complex interaction between the incident primary ion and a limited area of the solid surface in the immediate vicinity of the ion impact. The primary ion penetrates into the solid and transfers its energy to the crystal lattice. Part of this energy returns to the surface where, with sufficient transmission of energy to a surface particle, this will be emitted. The break-up of chemical compounds, especially ionic compounds, is particularly effective for ionization. When the primary ion energy is in the keV range, the secondary ions leave the surface with an average energy of several electronvolts. It follows that they can only originate from the uppermost monolayer of the solid, particularly if they are molecular ions (Figure 12).

The system for SIMS basically consists of four components: a primary ion source, a target chamber, a mass analyzer, and a detection system suitable for counting single ions. The removal of a surface layer by means of high primary ion current densities will establish whether a particular component is present only in the outermost molecular layer or whether it is also to be found in deeper layers. The most important characteristics of the static method of SIMS, the major feature of which is bombardment of a relatively large target area with an extremely low primary ion current density, can be summarized as follows:

1. The information depth is of the order of one monomolecular layer.
2. The emission of molecular ions enables chemical compounds to be determined.
3. Hydrogen and its compounds are detected.
4. Isotopes are distinguished.
5. The variation in ionization probability produces varying sensitivities for the different elements and compounds (a factor of approximately 10^3 for different metals).
6. For many elements and compounds, e.g., for metal oxides, inorganic and organic anion complexes, and hydrocarbon compounds, the detection limit is considerably below 1 ppm (10^{-6}) of a monolayer.
7. The depth distribution of a component in the region of the outermost monolayers can be determined by observing the time dependence of typical secondary ions.
8. When it is desired to follow changes in the chemical composition of an individual monolayer, with observation of single mass lines only, the total sputtering can be kept below 10^{-4} of a monolayer.

IV. SUMMARY

Since 1965 the SEM became available as a commercial instrument. In this short time the SEM has established itself as an exciting tool. Topographical studies and rapid X-ray microanalysis are in regular use. The SEM is capable of giving many signals that can all be electronically treated. For the future one can predict that by using image processing techniques, a total material characterization in a single instrument can be forseen.

The four analytical methods, EPMA, AES, ESCA, and SIMS, discussed in this paper have been developed and improved in recent years on the basis of well-known physical effects:

1. EPMA is based on the emission of characteristic X-rays from solids due to the impact of a high energetic electron beam.
2. AES is based on the emission of Auger electrons from solids, due to the impact of a low energetic electron beam.
3. The ESCA method is based on the emission of photoelectrons from solids, due to the impact of X-rays.
4. The SIMS method is based on the well-known fact that solid surfaces emit secondary ions under ion bombardment.

Property	EMPA	AES	ESCA	SIMS
Information depth	10,000 Å	15 Å	50 Å	1 monolayer
Detection of				
Elements	Yes	Yes	Yes	Yes
Isotopes	No	No	No	Yes
Compounds	No	(Chemical shift)	(Chemical shift)	Yes
Detection limit for elements	B (Z = 5)	Li (Z = 3)	Li (Z = 3)	H (Z = 1)
Sensitivity (g.cm^{-2})	10^{-7}	10^{-10}	10^{-9}	10^{-13}

Note:
- The EPMA is not suitable for monolayer analysis.
- The three methods, AES, ESCA, and SIMS, are all able to supply information on the composition of a solid surface, but vary in their average information depth.
- Elements are traced by all four methods.
- Isotopes are identified only by SIMS.
- With ESCA, chemical compounds can be identified through the chemical shift, and with SIMS they can be identified by means of the molecular ions emitted. With AES it is only in certain special cases that the chemical shift can be utilized for analysis.
- Hydrogen and its compounds can only be detected by the SIMS method.
- Sensitivity is high for all methods (better than 10^{-7} g/cm^2). The very high sensitivity of SIMS for many components has the disadvantage of a relatively large variation (of approximately three orders of magnitude) for some elements. This variation is considerably smaller with AES and ESCA.

The surface analysis methods have very low absolute detection limits, especially if one takes into consideration the dimensions of the surface area analyzed. Thus, they offer new possibilities for micro- and trace analysis in biomaterials science and engineering.

A brief reference should be made here to the necessary vacuum conditions. All the methods discussed require a sufficient mean free path for the participating photons, electrons, or ions leaving the surface to reach the detector. The vacuum necessary in individual cases will therefore lie between some 10^{-5} torr (the upper limit given by the mean free path requirement) and approximately 10^{-12} torr (e.g., for the observations of atomically clean surfaces).

REFERENCES

1. **Ducheyne, P., De Meester, P., Aernoudt, E., Martens, M., and Mulier, J. C.,** Fatigue fractures of the femoral component of Charnley and Charnley-Müller type total hip prostheses, *J. Biomed. Mater. Res.*, 9(4), 199, 1975.
2. **Ducheyne, P., Martens, M., De Meester, P., Aernoudt, E., and Mulier, J. C.,** The influence of a functional dynamic loading on bone ingrowth into surface pores of orthopaedic implants, *J. Biomed. Mater. Res.*, 11, 811, 1977.
3. **Ducheyne, P., De Meester, P., Aernoudt, E., Martens, M., and Mulier, J. C.,** Elastic and mechanical properties of porous fibre structures allowing bone ingrowth, in *Mechanical Properties of Biomaterials*, Hastings, G. W. and Williams, D. F., Eds., John Wiley & Sons, New York, 1980.
4. **Ducheyne, P., Martens, M., Aernoudt, E., Mulier, J. C., and De Meester, P.,** Skeletal fixation by metal fibre coating of the implant, *Acta Orthop. Belg.*, 10, 799, 1974.
5. **Ducheyne, P. and de Groot, K.,** In vivo surface activity of a hydroxyapatite alveolar bone substitute — a note, *J. Biomed. Mater. Res.*, 15, 441, 1981.
6. **Ducheyne, P., Hench, L. L., Kagan, A., Martens, M., and Mulier, J. C.,** Short term bonding behaviour of bioglass coatings on metal substrate, *Arch. Orthop. Traum. Surg.*, 94, 155, 1979.
7. **Ohnsorge, J. and Holm, R.,** Surface investigations of oxide layers on cobalt-chromium-alloyed orthopedic implants using ESCA techniques, *Med. Prog. Technol.*, 5, 171, 1978.

GENERAL REFERENCES

A. **Goldstein, J. I., Yakowitz, H., Newbury, D. E., Lifshin, E., Colby, J. W., and Coleman, J. R.,** *Practical Scanning Electron Microscopy*, Plenum Press, New York, 1975.
B. **Goldstein, J. I., Newbury, D. E., Echlin, P., Joy, D. C., Fiori, C., and Lifshin, E.,** *Scanning Electron Microscopy and X-ray Microanalysis*, Plenum Press, New York, 1981.
C. **McKindley, T. D., Heinrich, K. F. J., and Witty, D. B.,** The Electron Microprobe, *John Wiley & Sons*, New York, 1966.
D. **Lieng-Huang, L.,** *Characterisation of Metal and Polymer Surfaces*, Academic Press, New York, 1977.
E. **Ibach, H.,** *Electron Spectroscopy for Surface Analysis*, Springer-Verlag, Berlin, 1977.
F. **Carlson, T. A.,** *X-ray Photoelectron Spectroscopy*, Dowden, Hutchinson & Ross, Stroudsburg, Pa., 1978.
G. **Barr, Davis,** *Applied Surface Analysis*, ASTM Special Tech. Publ. 699, American Society for Testing and Materials, Philadelphia, 1980.

Chapter 4

MANUFACTURING PROCESSES OF METALS: THE PROCESSING AND PROPERTIES OF METAL IMPLANTS

Robert M. Pilliar

TABLE OF CONTENTS

I. INTRODUCTION

Metals by virtue of the nature of their interatomic bonds are characterized by a combination of strength and ductility that favors their use for load-bearing structures. The application of metals for surgical implants is restricted to a limited number of compositions that, in addition to being strong and ductile, are sufficiently corrosion resistant and biocompatible. However, failures of metallic implants in service have been reported and material-related causes for these failures have been suggested. The recent concern about such occurrences has acted as a stimulus for the introduction of new implant alloys and new processing methods for metal implant fabrication. The purpose of this article will be to review the metals presently used for orthopedic implants with specific reference to their processing and how it affects properties.

Prior to discussing the specifics of implant fabrication, a general description of the methods used will be presented. This will include descriptions of investment casting, mechanical forming, and powder metallurgical methods of powder consolidation by hot isostatic pressing (HIP).

II. METAL IMPLANT FABRICATION (GENERAL)

A. Investment Casting

Investment casting, also referred to as the lost wax process or precision casting, represents a method of forming metal shapes to almost finished dimensions. This process is used with metal alloys characterized by high work hardening coefficients for which mechanical forming would be difficult and costly, necessitating numerous intermediate heat treatments, and also for the formation of intricate shapes when machining would be difficult and uneconomical. It becomes a reasonably economic way of part fabrication when smaller numbers of parts are needed. As such it is used for surgical implant manufacture. Its advantage of allowing individual intricate shapes to be cast with close dimensional tolerance makes it suitable for use in prosthodontics.

The solidification of metal castings (to final shape or ingot form) can be considered to occur in two stages, namely, the nucleation and growth of solid phase from liquid. The final grain size of an as-cast ingot structure is determined by the rate of nucleation with a high nucleation rate resulting in many centers of solid-phase formation. These solid-phase regions will grow to the point of impingement, hence a fine grain size will result from a high nucleation rate. A fine-grained structure is desirable for better workability (if mechanical forming operations subsequently are used to achieve a final implant shape — not the case for investment cast parts) and for better mechanical properties (higher tensile and fatigue strengths, greater ductility). In order to achieve a high nucleation rate during solidification, rapid cooling rates and the introduction of nucleating agents are employed. The rapid cooling of the molten metal results in increased supercooling of the liquid phase prior to formation of stable nuclei. The increased supercooling promotes the formation of solid-phase nuclei and also the lower temperatures at which they form limits their subsequent growth rates. The large number of nuclei and their lower growth rates (dependent on solid-state diffusion) result in finer-grained structures. The use of nucleating agents affects the nucleation process resulting in finer-grained structures.

Increased supercooling as achieved by rapid cooling of the melt is normally limited by the requirement that the liquid phase flow readily within a casting mold. Therefore, to achieve effective filling of an intricate mold by liquid phase, the molten metal is heated to a temperature in excess of its melting point prior to pouring. Additionally, the casting mold is normally preheated. The result of this liquid superheating and mold preheating in addition to improved flow characteristics of the liquid melt is a reduction in the rate of cooling and hence, structures after casting are normally coarse grained. Nucleating agents are used to promote more rapid nucleation under these conditions. Several studies have been reported on autocatalytic nucleation using ultrasound and other mixing methods to cause breakup of the initially formed solid phase. The resulting fine solid particles act as nuclei to promote further solidification, resulting in more homogeneous nucleation and finer-grained structures. These methods have yet to find practical use in implant fabrication.

During alloy casting, the solidification process is accompanied by a partitioning of component elements as determined by the alloy's phase diagram. Thus, a higher melting temperature composition will solidify first and this normally corresponds to a solute-poor composition. There follows a continuous enrichment of solute at the advancing solid-liquid interface resulting in a solute-rich composition solidifying last. The resulting as-cast structure is, therefore, very heterogeneous in composition with a fairly regular variation in composition throughout the casting corresponding to the initially solidifying solute-depleted zones and the solute-enriched interdentritic regions (a dendritic morphology is common during metal solidification). This as-cast structure is referred to as a "cored" structure and normally is not acceptable.

Excess solute enrichment can cause the formation of secondary phases with poor properties, thereby limiting the overall cast part properties. For this reason procedures are normally taken after casting to homogenize the structure. Although this can be achieved by elevated temperature diffusion annealing alone, the times to obtain homogeneity are long and normally castings are mechanically worked prior to annealing. The mechanical working decreases diffusion distances as well as increasing diffusion rates (through increased numbers of atomic lattice defects). For investment cast parts, mechanical working is not practical and, therefore, although homogeneous compositions are possible with sufficiently long diffusion anneals, chemical heterogeneities often persist in the final product.

The procedure of investment casting is unique only in the method of mold formation. Whereas conventional casting relies on pouring molten metal into a retrievable and reusable (sand) mold, investment casting utilizes a refractory ceramic shell mold that is used only once. The process consists of forming a wax replica of the part to be cast, either by hand or by injection of wax into a metal die. The wax replica (and others like it) is normally attached to a common feeding sprue with appropriate risers to allow for efficient filling during pouring of the liquid metal. This multipart wax pattern is commonly referred to as a "tree". Design of the casting tree remains an art for both proper liquid flow and for the achievement of desired cooling rates at critical regions of the casting.

After formation, the wax "tree" is dipped into a slurry of a refractory metal. Typically, this first slurry coating consists of a suspension of fine silica particles in an ethyl silicate solution. Since this initial coating represents the future inner surface of the casting mold, it commonly has dispersed through it particles that will serve as nucleating agents during the solidification process. A number of first coat slurry compositions are used presently.[1] To provide strength to the investment, this first coating then is allowed to dry and subsequent coarser silica particle-containing slurries are applied to and dried on the wax patterns. The final thickness of the refractory mold must provide sufficient strength while allowing as rapid a cooling rate as possible.

The refractory-coated wax pattern after drying is heated to a temperature typically in the range of 95 to 150°C. At this time the wax becomes molten and runs out, leaving the investment mold. The fine particle composition of the inner coating of the investment results in a relatively smooth surface in the shell mold. This mold then is preheated to a higher temperature (the mold preheat temperature). This results in mold out-gassing, removal of any trace amounts of wax, complete curing of the mold, and the mold preheating that promotes liquid metal flow into all parts of the mold on casting.

The casting operation itself is achieved by pouring the molten metal into the mold either in air or under a protective atmosphere to avoid oxidation and contamination of the casting.

In addition, mold filling can be promoted by use of centrifugal force or gas pressure. Gravity feed methods also are used and, provided melt and mold temperatures and mold designs are suitable, will result in sound castings.

In addition to the coarser structures that result during casting because of the limitations on cooling rates, defective castings can result from porosity within the cast parts. This porosity can be gross and extremely detrimental to mechanical properties rendering the parts unusable or it can consist of a distribution of fine pores that can be tolerated (and, in fact, cannot be avoided). Examples of porous structures resulting on castings are shown in Figure 1 (A and B). This porosity can result from

1. Gases liberated during the solidification process, since the solubility of gases is greater in the liquid phase than in the solid phase (Figure 1A)
2. Shrinkage which occurs on solidification and is normally interdendritic (Figure 1B)
3. Poor liquid flow to regions of the mold

Gas porosity can be reduced by use of vacuum cast melt stock. Voids due to poor liquid flow can be avoided by the use of proper pouring conditions and mold design. The interdendritic shrinkage porosity, however, is intrinsic to the solidifying metal. Its size and distribution can be affected by solidification conditions with fine structures yielding smaller and more frequent pores. This is normally preferable since these finer pores can be less significant in causing regions of stress intensification and, therefore, when compared to coarser pores, they are less effective as crack initiators during subsequent forming operations or in-service loading of the part.

Internal voids, particularly those resulting from solid phase shrinkage, can be eliminated by HIP of the cast parts. This represents a process by which parts are subjected to isostatic pressures at elevated temperatures for prolonged periods of time. The pressures, temperatures, and times are so chosen that metal deformation corresponding to high temperature creep will occur resulting in eventual closure of internal voids.

As well as void defects, nonmetallic inclusions can be found in castings when either poor quality melt stock is used or when refractory particles from melt crucibles or casting molds are inadvertently mixed into the melt. This type of defect cannot be eliminated once present, and again, depending on particle size and distribution, it can cause significant deterioration in implant properties.

B. Mechanical Forming of Parts

Because of the lower mechanical properties of as-cast alloys, critical, load-bearing, engineering structures are normally made from wrought alloys. The process used for forming wrought metal products involves the mechanical deformation of cast material in order to refine grain structure, break up brittle secondary phase regions, close internal porosity, and, through the introduction of internal strain energy, promote chemical diffusion upon subsequent annealing to achieve more uniform compositions. A number of different mechanical working operations are possible and these can be performed either hot or cold and have a profound influence on structures and properties.

Normally, a cast ingot will be hot forged, rolled, or extruded to a shape and supplied in bar, rod, or plate form for further processing. This stock can then be worked further by rolling, forging, extruding, swaging, and/or drawing to a final shape. The process and conditions used affect part properties and this and economic considerations dictate the procedure used.

Mechanical working to shape a part requires plastic flow of the metal. Thus, the material's yield stress will determine the forces necessary to achieve a shape change. Cold working (deformation below the recrystallization temperature) results in structural changes that lead

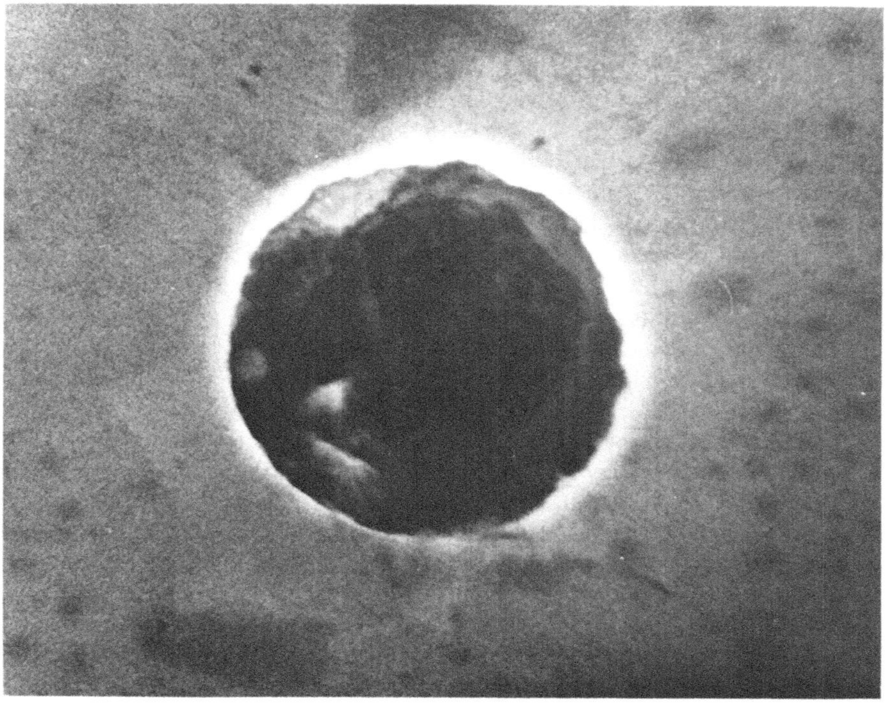

A

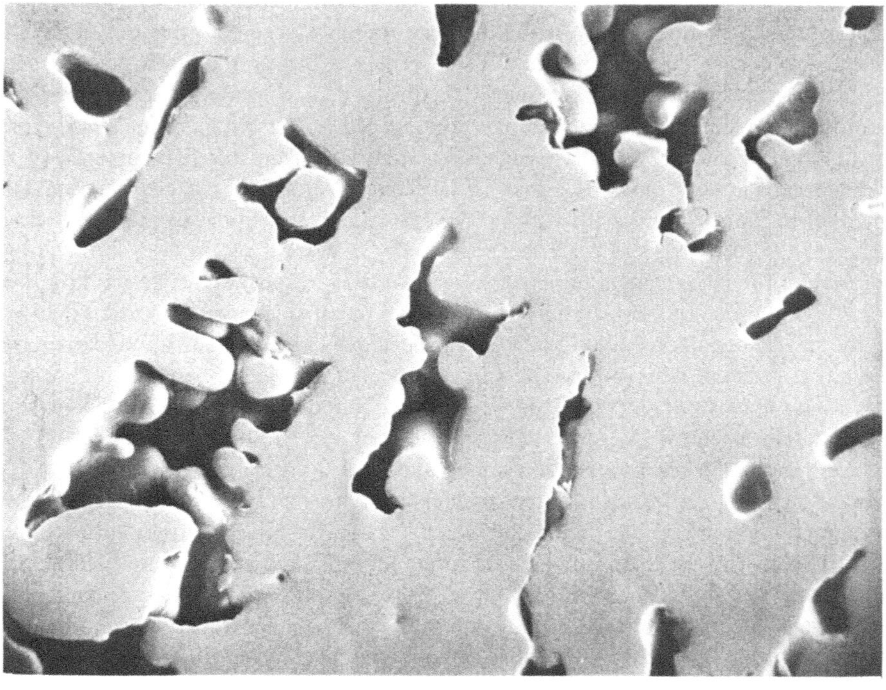

B

FIGURE 1. Casting voids in a cast Co-Cr-Mo implant; scanning electron micrographs of polished as-cast samples. (A) Gas void (magnification × 1000); (B) interdentritic shrinkage porosity (magnification × 120).

yield stress will determine the forces necessary to achieve a shape change. Cold working (deformation below the recrystallization temperature) results in structural changes that lead to increased yield strengths and reduced ductilities. The reduction in ductility often limits the extent of cold working so that to complete a required shaping operation intermediate anneals are necessary. These intermediate anneals result in reduction of internal strain energy through processes of recovery and recrystallization (atomic rearrangements occurring on annealing). As a result, yield strengths and ductilities return to the values exhibited by the annealed, nondeformed material. Excessive cold working of a material can cause cracking to occur as a result of excessive local plastic strains. Such difficulties associated with cold working lead to the common use of elevated temperature forming and shaping methods.

A higher temperature results in a decreased yield stress and increased ductility. Thus, hot working of metals can be achieved with less force and with less risk of cracking or tearing. The hot working temperature must be high enough that the recovery process can occur at a rate greater than the work hardening process. This defines a lower working temperature for a given pressure and deformation rate to achieve a specific amount of deformation. During mechanical working, strain energy is converted partly to thermal energy. If the strain rate is high enough, the work piece will increase in temperature during the working process. This effect must be taken into account in determining the maximum working temperature since this thermal energy can result in localized melting and this in turn can cause part cracking during forming, a process referred to as hot shortness. In general, hot working temperatures are chosen as low as possible for economic reasons. The structure and properties of hot worked metals are strongly influenced by the temperature at which deformation is completed and by the rate of cooling of the part from this temperature. These factors will influence grain size, second phase distribution, and state of internal strain in the finished part.

To completely convert an as-cast structure to a high strength, wrought structure requires at least a 70% reduction in area.[2] To avoid the need for such large deformations in the final forming operation, cast ingots are normally worked to a bar or rod form and then in a separate operation are worked to final shape. The direction of working influences properties and, therefore, to ensure a satisfactory final working process, it is important that the prior working direction is known. This is particularly important if anisotropic properties result since orientation of a work piece such that on subsequent deformation excessive elongation occurs parallel to a prior transverse working direction can result in the formation of cracks in the material.

The use of a finish anneal after a working operation (especially cold working) will determine the properties of the material. Thus, if high ductility is more important than high strength, the finish anneal temperature will be chosen above the recrystallization temperature. To ensure a fine-grained, recrystallized structure with a combination of strength and ductility high nucleation rates of the recrystallized grains are required. This normally is achieved by heavy deformation at below the recrystallization temperature, followed by anneals at temperatures well above the recrystallization temperature for short times.

Mechanical working can result in heterogeneous structures and properties. The degree of deformation and cooling rate after annealing can vary significantly from region to region in large parts. This can result in significant variations in structures and properties throughout the part. These variations can be beneficial as, for example, when a fine-grained, more fatigue-resistant structure is formed at the outer surface. It can, however, cause difficulties in subsequent processing operations because of nonhomogeneous deformation and cracking. One method of avoiding heterogeneous structures that can cause such nonhomogeneous deformation is to use metal powders to form fully dense parts.

1. Powder Metal Part Manufacture

Metal alloy powders are formed by atomization, a process by which molten metal is

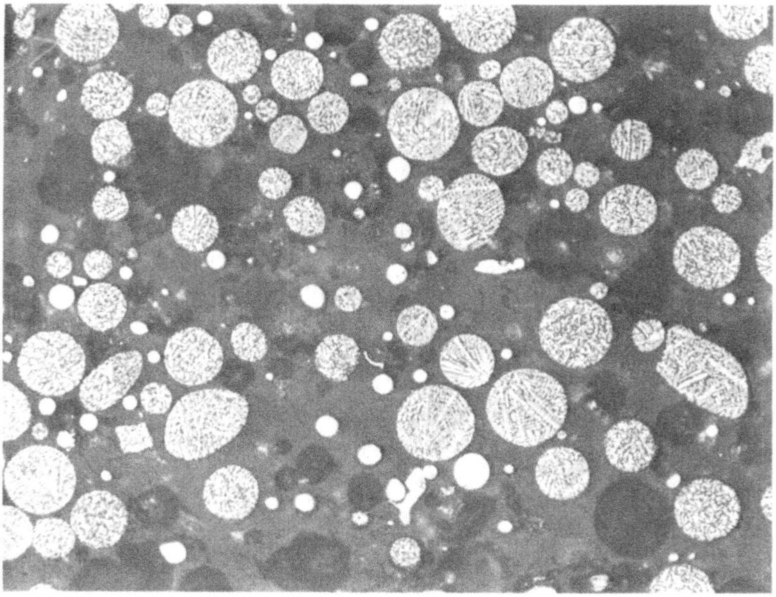

FIGURE 2. Argon gas atomized Co-Cr-Mo alloy powders. Note the spherical particle shapes resulting from the relatively slow cooling rates of the gas atomized powders and the fine-cored structure within the individual particles. (Magnification × 250.)

dispersed into fine particles that subsequently solidify. The dispersion can be achieved by a high pressure inert gas jet (inert gas atomization) or by a water jet (water atomization) acting on the molten metal stream during pouring or by centrifugal forces acting on a consumable electrode during its arc melting (spun electrode process). The atomization process parameters (e.g., gas or water pressure, consumable electrode spinning velocity, melt temperature) can be controlled to give desired powder particle size and distribution. The rapid solidification rates of the small particles that occur during atomization result in fine microstructures within these ''minicasting'' particles. The shape and structure of Co-Cr-Mo alloy powders formed by inert gas atomization are shown in Figure 2. Retention of this fine structure to the final consolidated form results in materials of cast composition exhibiting higher strengths and ductilities than conventional cast alloys.

Although formation of final shapes from metal powders is possible, the so-called net shape process, the process of forming full density final forms is expensive and economically not feasible at present. An alternate method of forming full density powder-made parts is to shape by conventional forging processes powder metal-made, full-density bar or rod stock. The fine microstructure and resulting good formability of the powder-made, full-density starting material facilitates its forging.

The consolidation of the metal powder can be achieved by various hot deformation processes. To avoid contamination and oxidation of the metal powders, they are normally placed in evacuated containers and sealed prior to consolidation. Typically, clean metal powders are placed in stainless steel containers that are evacuated to about 10^{-6} torr and sealed. The sealed, powder-containing containers are subsequently hot worked in order to consolidate the powders. This hot working can be achieved by hot extrusion, forging, or isostatic pressing. After this, the deformed stainless steel container material is removed (by machining or chemical pickling) and the resulting full-density metal stock is processed to final shape. Critical to the formation of high-quality, full-density material is the surface characteristic of the atomized powders (both prior to and during consolidation). Powder cleanliness is critical in obtaining high toughness and high fatigue strength in the fully densified material.[3,4]

III. IMPLANT ALLOYS

A. Co-Base Alloys

1. Co-Cr-Mo

The manufacture of artificial skeletal joint replacements by investment casting is particularly attractive because of the complex shapes that often are encountered. Although grinding, honing, and polishing operations are used for forming the high finish articulating surfaces, the use of investment casting minimizes the amount of machining necessary.

The compositional specifications for the cast Co-base implant alloys are shown in Table 1. These alloys are cast at temperatures of 1350 to 1450°C, depending on their exact compositions.[5] A variation in Ni content of, for example, 0.5 to 2.5 wt %, results in a lower melting temperature for the higher Ni-containing alloys that supposedly also results in improved flow characteristics during casting.

Figure 3 shows a typical as-cast Co-Cr-Mo alloy structure. It is highly cored with the interdendritic zones enriched in Mo, Cr, and C. Various studies have been made to identify the phases present in the as-cast structure. Kilner et al.,[6] utilizing electron and X-ray diffraction techniques and microprobe analysis of the interdentritic regions, identified a Cr-rich $M_{23}C_6$ phase, sigma phase (also rich in Cr and Mo), and a gamma matrix phase within these zones. The brittle nature of the carbide and intermetallic phases and their shapes and distribution are responsible for the limited ductility (typically 5 to 8% elongation only)[7] of the as-cast alloy. The heterogeneous composition of the as-cast structure also has been reported to adversely affect corrosion resistance.[8,9] Because of these factors load-bearing cast Co-Cr-Mo alloy implants such as femoral hip components are normally given a high temperature anneal. This heat treatment typically consists of a 1220 to 1230°C, 1-hr treatment followed by rapid cooling to room temperature, and its effect is partial homogenization of the as-cast, cored structure and dissolution of the $M_{23}C_6$ carbide and sigma intermetallic phases in the gamma matrix phase. Figure 4 shows a structure typical of a 1230°C, 1-hr annealed sample. This treatment results in increased ductility as indicated in Table 2. At temperatures just below 1230°C, prolonged heat treatment (>24 hr) results in the complete solutionizing of the $M_{23}C_6$ and intermetallic phases. It was shown by Clemow and Daniell that a 64-hr, 1212°C treatment also resulted in complete homogenization of the alloy.[10] The presence of M_6C in the as-cast and partially homogenized materials was suggested in their study. Kilner et al.[6] were not able to conclusively identify the presence of M_6C in their as-cast and annealed samples.

The presence of a eutectic melting point at about 1235°C has been inferred in both the studies by Kilner et al.[6] and by Clemow and Daniells.[10] Incipient melting of the interdendritic, solute-enriched regions above 1235°C can lead to liquid phase formation along grain boundaries, thereby leaving a continuous brittle phase stringered throughout the structure. Further, the presence of these zones at the surface of the sample may result in regions prone to easy fatigue crack initiation.[11]

The possible decrease in fatigue strengths as a result of too high a solution anneal temperature exemplifies the importance of fully appreciating the effect of processing on structure and properties of these alloys. Rather than increasing the diffusion rate and sample homogenization, the higher annealing temperature can result in mechanical property deterioration of the material.

As noted previously, casting voids can form in this alloy and if present in critical regions of the implant can significantly affect fatigue properties. A number of instances of void-related fatigue failures of cast Co-Cr-Mo implants have been reported.[12] The harmful effect

Table 1
SURGICAL IMPLANT ALLOY COMPOSITIONS (WEIGHT PERCENT)

	Co-base alloys				Ferrous alloys			
					ASTM F138,F139		ASTM F55	
Element	Cast (ASTM F75)	Wrought (ASTM F90)	Syntacoben® (ASTM F563)	Co-Ni alloys MP35N	A	B	A	B
C	0.35 max	0.05—0.15	0.0005 max	0.025 max	0.03 max	0.08 max	0.03 max	0.08 max
Co	Balance	Balance	Balance	Balance	—	—	—	—
Cr	27.0—30.0	19.0—21.0	18.0—22.0	19.0—21.0	17.00	20.00	17.00	20.00
Fe	0.75 max	3.0 max	4.00—6.0	1.0 max	Balance		Balance	
Mo	5.0—7.0	—	3.0—4.0	9.0—10.5	2.00	4.00	2.00	4.00
Mn	1.00 max	2.00 max	1.0 max	0.15 max	2.00 max		2.00 max	
Ni	2.5 max	9.0—11.0	15.0—25.0	33.0—37.0	10.00	14.00	10.00	14.00
P	—	—	—	—	0.03 max		0.025 max	
S	—	—	0.010 max	0.010 max	0.03 max		0.010 max	
Si	1.00 max	1.00 max	0.50 max	0.015 max	0.75 max		0.75 max	
Ti	—	—	0.5—3.5	—	—	—	—	—

Titanium alloys

Element	C.P. Titanium (ASTM F67)	Ti6A14V
H	0.0125—0.015	0.0125 max
C	0.10 max	0.08 max
O	0.18—0.40	0.13 max
N	0.03—0.05	0.05 max
Al	—	5.5—6.5
V	—	3.5—4.5
Fe	0.20—0.50	0.25 max
Ti	Balance	Balance

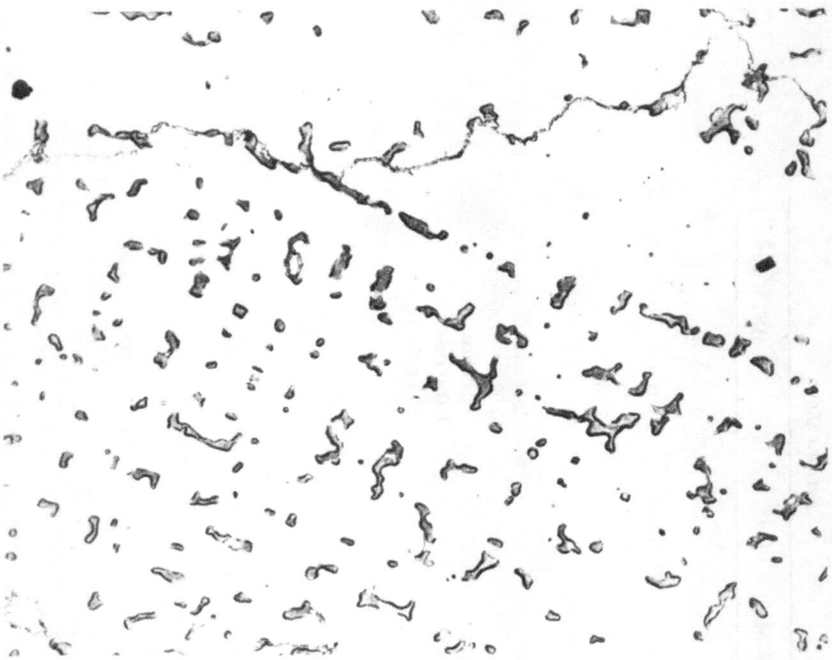

FIGURE 3. Microstructure of an as-cast Co-Cr-Mo implant alloy. Note the pronounced cored structure. (Magnification × 150.)

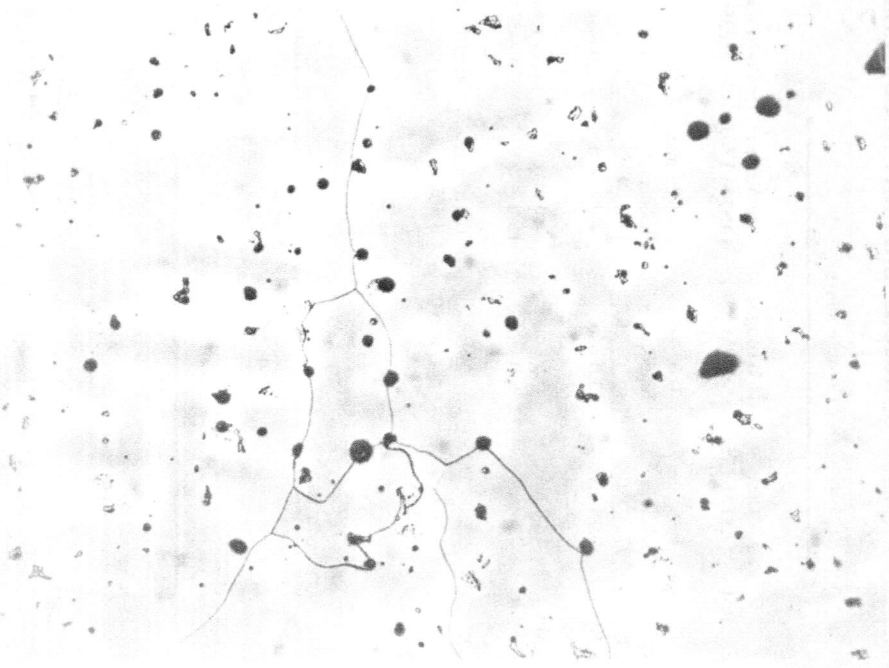

FIGURE 4. Microstructure of a solution annealed (1230°C, 1 hr) Co-Cr-Mo implant. This represents the conventional treatment of a cast Co-base implant. Significant compositional heterogeneity persists. (Magnification × 190.)

Table 2
MECHANICAL PROPERTIES OF THE Co-Cr-Mo SURGICAL IMPLANT ALLOYS

Treatment	σ_y(MPa)	σ_{ult} (MPa)	% Elongation	σ_{end} (MPa) (10^7 c; R = −1)	Ref.
As-cast (F-75-67)	430—490	716—890	5—8	300	8
Solution annealed (S.A.) (1230°C; 1 hr, water quench)	450—492	731—889	11—17	250	8
S.A. + aged 650°C/20 hr	444—509	747—952	10—13.5	—	8
Müller-cast alloy	600	1000	25	400	14
815°C/4 hr + 1225°C/4 hr; brine quenched	525	1100	24	—	15
Cast and extruded @ 1200°C; annealed 1100°C/2 hr	731	945	17	345	18
Cast and hot forged and rolled @ 1175°C + cold rolled 10% + 1050°C/40 min; air cooled	876	1360	19	—	18
Modified low C-alloy treated as above	690	1640	26	670	18
FHS vitallium	890—1028	1408—1511	28	793—966	20
Micrograin p/m[a], HIP	841	1277	14	725	22

[a] p/m = powder metallurgy.

of these voids can be minimized by choosing casting parameters that will result in very fine shrinkage voids only. This can be achieved by rapid solidification rates using minimum melt temperatures and nucleating agents incorporated in the investment mould.

The use of HIP of as-cast, Co-based alloy implants has been advocated to eliminate microporosity, thereby causing improvements in mechanical properties.[7-9] Hodge and Lee treated cast samples by HIP (1230°C, 105 MPa, 2 hr, presumably in argon), followed by a normal solution anneal treatment.[8] They reported an increase in reduction in area of from 5 to 17% for tensile specimens so treated, although yield and ultimate strengths were not greatly affected. They also reported a large improvement in fatigue life after such a treatment. The solution anneal subsequent to HIPing was necessary because of the unavoidable slow cooling rates in the hot isostatic press. Exposure of specimens to temperature ranges where $M_{23}C_6$ is stable and diffusion rates are high (around 1100°C) will result in the formation of lamellar $M_{23}C_6$ grain boundary precipitates. These are known to be detrimental to fatigue properties, and, therefore, Hodge and Lee gave their samples a 1230°C, 1-hr solution anneal after HIP, followed by rapid cooling to room temperature. It should be noted, however, that the HIP treatment can only cause healing of internal voids, and any surface defects, rather than being closed, will oxidize and poorer properties will result.[9] Furthermore, the usefulness of HIP treatment for healing macrovoids is limited and inclusions, of course, are not eliminated. Hence, a medium-quality casting, containing internal casting voids, can be improved by HIP, whereas a poor casting containing surface defects and inclusions is unlikely to be improved by this treatment.

The effect of aging treatments for Co-Cr-Mo alloys following solution annealing has been studied by a number of workers.[7,9,13] Weeton and Signorelli[13] studied the effect of (1) isothermal treatments following a 1230°C, 72-hr solution anneal and (2) aging after solution annealing and quenching of a hot rolled, cast Co-Cr-Mo alloy. The results of the isothermally treated samples indicated the formation of lamellar $M_{23}C_6$ after 1065, 955, and 815°C anneals, while at the lower temperatures (955 and 815°C), oriented, intragranular precipitates formed.

The aging treatments resulted in precipitates forming along what were identified as slip lines and twin boundaries and, in fact, were probably transformed phase regions (see below). It was shown that hardness increased appreciably with aging above 650°C, with peak hardness being reached by aging for 72 hr at 760 to 815°C. Although these early studies were stimulated by a search for a high-temperature, creep-resistant Co-base alloy for turbine blade applications, the findings proved useful in indicating the response of the Co-Cr-Mo cast implant alloys to heat treatment.

Based on these earlier studies Hollander and Wulff[9] investigated the effect of a 650°C, 20-hr aging treatment after solution annealing and quenching of a cast Co-Cr-Mo implant alloy. The effect of this aging treatment was a slight increase in ultimate and yield strengths for some as-cast materials. However, lower properties were observed after this treatment for implant alloys made by some manufacturers.

The Co-Cr-Mo alloy is face centered cubic (fcc) at temperatures above about 890°C. The presence of other phases in conventionally cast and solution-annealed implants has been discussed and is related to the heterogeneous composition resulting from casting and its partial retention after a 1-hr, 1230°C, solution anneal. For a completely homogenized alloy, below 890°C the stable allotrope is hexagonal close packed (hcp). The transformation from fcc to hcp is sluggish, however, so that on normal cooling (air or water quench), the fcc phase is retained. However, on subsequent annealing at temperatures where the hcp phase is stable (<890°C), bands of hcp phase will form. These hcp bands throughout the metastable fcc matrix provide sites for precipitation of $M_{23}C_6$. Tensile strengths and ductilities are dependent on this aging treatment. (Table 2).

Much higher mechanical properties for a cast Co-Cr-Mo alloy have been reported by Müller.[14] The material, referred to as Endocast, is characterized by higher strengths and ductilities as indicated in Table 2. Metallographic studies of this material by Cohen et al.[15] indicated the presence of intragranular carbonitrides and sigma-phase precipitates. It was postulated by these workers that these precipitates were of a size to act as stacking fault generators during quenching of the alloys and of sufficient volume percent to cause a significant improvement in mechanical properties. Strains due to differential thermal expansion of precipitate and matrix phase have been reported to generate dislocations in other alloys[16] and presumably a similar mechanism could act for the Co-Cr-Mo alloy. The alloy was shown to have a higher nitrogen content (0.2 wt %) compared to conventional Co-Cr-Mo implant alloys.[15] Cohen et al. reported that a heat treatment schedule consisting of an 815°C, 4-hr aging anneal, followed by a 1225°C, 4-hr solution anneal, followed by a rapid quench, resulted in improved properties (Table 2). It is difficult to rationalize the improved properties because of this treatment. Attempts to duplicate these results by the author and co-workers have failed to produce the high-strength material. Cohen et al. also reported on the effect of minor additions (about 0.1 wt %) of Al, B, Nb, Ta, Ti, and Zr in cast Co-Cr-Mo alloys to further improve strengths and ductilities. Their studies indicated the benefits of a 750°C, 2-hr or 815°C, 1/2-hr aging anneal subsequent to the treatment described above for further improvements in strength. This final aging anneal resulted in carbide precipitation in the hcp bands that formed in the highly faulted and metastable fcc structure.

The studies reported by Cohen et al.[15] and the studies by Müller[14] indicate significant property improvements for cast Co-Cr-Mo alloys by thermal treatments alone. Alternatively, mechanical properties of these alloys can be improved using thermomechanical treatments. Weeton and Signorelli[13] in the mid 1950s investigated hot worked Co-Cr-Mo alloys. Devine et al.[17] further studied the corrosion and mechanical properties of wrought Co-Cr-Mo alloys in the early 1970s. Hot forging or extrusion of the cast ingots is possible at about 1200°C. The effect of fragmentation of the large, as-cast carbide and intermetallic networks while working at these higher temperatures results in a more workable alloy that subsequently can be formed to a final implant shape. Devine and Wulff[18] compared the cast and wrought

HS21 alloys and reported significant increases in strength and ductility and improved corrosion resistance for the wrought Co-Cr-Mo alloy compared to the cast and solution annealed alloy — Table 2. For an alloy hot forged and rolled at 1175°C, followed by a 10% cold rolling reduction and annealed at 1050°C for 40 min and air cooled, increased tensile properties were reported (Table 2). A lower C-containing, modified HS21 alloy also was studied. By reducing carbon content to 0.145 w/o, the hot and cold working characteristics of the alloy were improved while retaining the good mechanical properties (Table 2). Fatigue studies of this lower carbon, wrought alloy demonstrated a dramatic increase in fatigue strength compared to the cast and solution annealed conventional Co-Cr-Mo alloy.[18] For specimens loaded in reverse bending (stress ratio, $R = -1$), the 10^7 c endurance limit was nearly doubled (end = 676 MPa, compared to 350 MPa). The improved mechanical properties and formability of the wrought Co-Cr-Mo alloy was related to its much finer grain size and the more uniform distribution of finer carbides that resulted from the initial hot working operation. An additional consequence of this processing was a more homogeneous alloy composition. The studies by Devine and Wulff also indicated improved corrosion properties for this more homogeneous structure.

Van der Sande et al.[19] investigated the strengthening mechanisms operating in Co-Cr-Mo alloys (both for the cast, solution annealed, and aged and for the modified, low carbon wrought alloy after solution treatment and aging). The strengthening resulting from aging at 650 to 750°C was attributed to faulted fcc regions and to martensitic hcp bands that formed within the metastable fcc phase. Further strengthening occurred as a result of $M_{23}C_6$ precipitates forming in the hcp transformed zones. The final properties were related to the density of stacking faults, the amount of hcp transformed phase, and the extent of carbide precipitation. Van der Sande et al. showed with their transmission electron microscope studies that quenching both the normal (higher C) cast and solution annealed (1230°C, 1 hr), Co-Cr-Mo alloy and the wrought, modified Co-Cr-Mo alloy solution annealed at 1230°C for 4 hr resulted in the formation of fcc stacking faults and some martensitic transformation of the fcc phase to a heavily faulted hcp structure. For the cast material, longer aging times at 650°C resulted in a higher proportion of martensitic hcp phase with approximately 90% hcp phase resulting after a 50-hr aging treatment. Precipitation of $M_{23}C_6$ also was observed to increase with aging time. The wrought, modified (low carbon) alloy aged at 650°C showed an increasing density of fcc stacking faults on planes (111) with increasing aging time. The martensitic transformation to hcp also increased with aging time. For this lower carbon material, however, no carbide precipitation in these hcp zones was observed after aging at 650°C. A 750°C anneal accelerated the formation of the hcp phase. This higher aging temperature resulted in the formation of hcp phase and carbide precipitation was observed in these regions.

Low carbon, forged, Co-Cr-Mo alloy implant components are presently manufactured.[20] In addition to a much lower carbon content (0.05% by weight), the chromium composition is somewhat lower (Cr equals 26.5%, compared to 27 to 30% for the conventional, cast alloy — ASTM* F75). These composition modifications are intended to enhance implant forgeability. As a result of the forging operation a fine-grained, fibered microstructure results. The fiber structure orientation corresponds to the direction of metal flow during the forging operation. Although the processing details have not been disclosed, hot forging of wrought Co-Cr-Mo bar stock with the finish forging temperature below the recrystallization temperature of the alloy would result in the type of microstructure observed in these low carbon, forged Co-Cr-Mo implants. For these implants, the grain size is observed to vary across the implant section. At the outer peripheral zone, a grain size of ASTM No. 8 or finer results, while in the inner zone a grain size of No. 5 or finer is observed. This grain size variation also occurs from the coarser-grained proximal end of the stem to the finer-grained distal tip. This grain size variation throughout the part suggests limited working of the material

* American Society for Testing and Materials, Philadelphia, Pa.

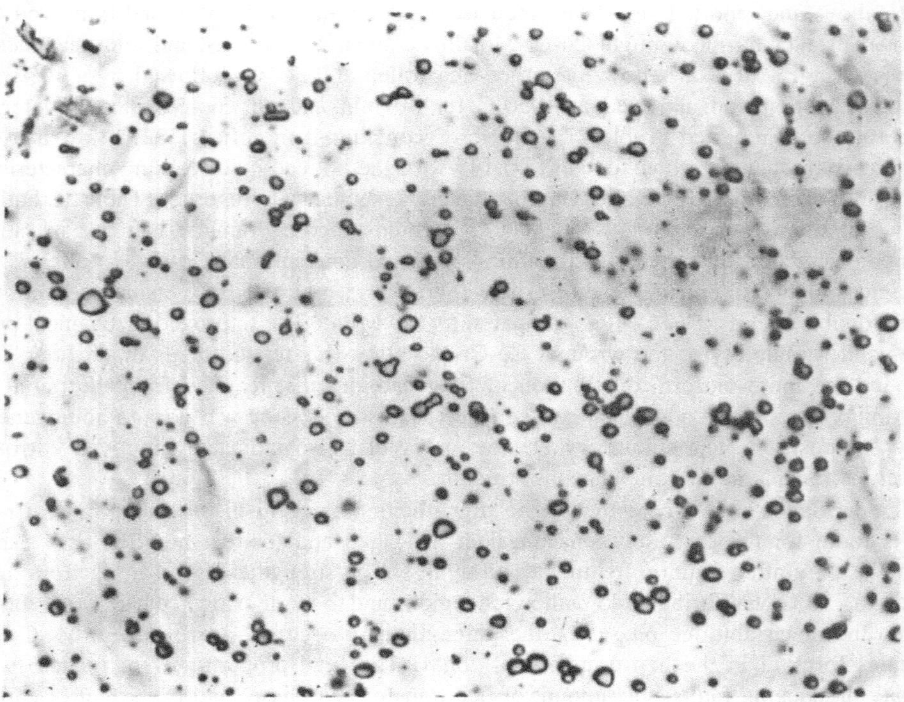

FIGURE 5. Powder-made, hot isostatically pressed Co-Cr-Mo alloy implant. (Magnification × 950.)

at below the recrystallization temperature during the forging operation. Comparison of the mechanical properties of implants so formed and those reported by Devine and Wulff[18] for modified Co-Cr-Mo alloy hot worked at 1150°C followed by limited cold working (10% reduction in area) and heat treatment at 1065°C for 30 min followed by air cooling indicates similar properties (Table 2).

The development of powder metal-made Co-Cr-Mo alloy implants has been referred to previously. By HIP of Co-Cr-Mo atomized alloy powder to full density, a very fine structured material can be formed.[21] Figure 5 shows a structure typical of a powder-made material corresponding to consolidation by HIP at 1100°C, 100 MN-m^{-2} for 1 hr. Both a very fine grain size and a fine and uniform secondary phase distribution is seen. This fine microstructure allows easy subsequent shaping of the material by forging, and by performing this final forging operation at a sufficiently low temperature, excessive grain growth and secondary phase coarsening can be avoided. As a result, implants made from Co-Cr-Mo alloy with high mechanical properties are formed (Table 2). The corrosion resistance of these powder-made parts is reportedly similar to that of conventionally cast and solution annealed Co-Cr-Mo alloy.[22]

2. *Co-Cr-W-Ni*

Because of the limited mechanical properties of the conventional cast and annealed Co-Cr-Mo alloys, a higher strength, wrought Co-Cr-W-Ni alloy (see Table 1 for chemical composition) was introduced for implant fabrication in the 1960s (ASTM F90). This alloy still is used for fabricating some implant shapes. The mechanical properties of the material are shown in Table 3. It differs from the cast Co-Cr-Mo alloy in having a much lower rate of work hardening, and hence, reductions of up to 40% are readily achieved on cold working. Microstructures of a Co-Cr-W-Ni implant alloy in the cold worked and annealed condition

Table 3
MECHANICAL PROPERTIES OF THE Co-Cr-W-Ni SURGICAL IMPLANT ALLOY[20]

Treatment	σ_y (MPa)	σ_{ult} (MPa)	% Elongation	σ_{end} (10^7 c; R = −1)	Ref.
Annealed	650	1220	37	—	23
Annealed	450	950	60	—	20
Cold worked (17.5% R.A.)[a]	1180	1350	22	490	23
Cold worked (44% R.A.)[a]	1610	1900	10	587	23
Cold worked	1300	1520	—	—	20
Syntacoben®,-annealed	275	600	50	—	24
Syntacoben®,-cold worked	828	1000	18	—	24

[a] R.A. = reduction in area.

are shown in Figures 6A and B. Hughes and Lane studied the effects of heat treatment and deformation on the properties of this alloy with particular consideration being given to its use for orthopedic implants.[23] They studied the dependence of strength and ductility on the degree of cold working and heat treatment. It was shown that aging at 650°C after cold working resulted in higher strengths with a corresponding reduction in ductility. Although the wrought Co-Cr-W-Ni alloy (HS25) exhibits superior mechanical properties compared to the cast and heat treated Co-Cr-Mo alloy, a question remains as to its susceptibility to crevice corrosion in the cold-worked condition. The Co-Cr-W-Ni alloy is widely used for temporary implants such as fracture fixation plates, intramedullary rods, and nails. For these applications, the ductility of the material combined with its high strength and fatigue resistance are ideal since custom bending of the implant in the operating room is a desirable feature.

A more recent material introduced for similar applications (i.e., fracture fixation plates, intramedullary rods) is a wrought Co-Ni-Cr-Mo-W-Fe alloy (trade name, Syntacoben®[24] — Table 1). It is claimed that this material possesses excellent corrosion resistance (not susceptible to pitting or crevice corrosion under simulated body conditions). The reported mechanical properties for various degrees of cold working and aging are included in Table 3. Fatigue studies of the alloy cold worked 30% indicate a 10^7 endurance limit of approximately 530 MPa (R = −1).[25]

B. Co-Ni-Base Alloys

1. MP35N (Co-Ni-Cr-Mo)

Orthopedic implants fabricated from MP35N (multiphase Co-Ni-base alloy of nominal composition, 35 Co, 35 Ni, 20 Cr, 10 Mo, Table 1) have been available for the past decade. The alloy is used in the wrought form typically being hot forged or rolled at temperatures up to 1125°C.[26,27] Deformation of the as-cast ingot below 870°C is not recommended since this lower temperature working can lead to edge cracking. However, once the as-cast structure has been broken down by hot working, the alloy can be further worked both hot and cold, and, in fact, the development of high strengths depends on lower-temperature deformation. Although the alloy is fcc at higher temperatures and remains so on slow cooling to room temperature, the equilibrium phase structure below about 425°C is hcp. The sluggish nature of the transformation from fcc to hcp prevents the formation of this phase on cooling alone. However, mechanical deformation below 425°C induces the transformation with narrow hcp bands forming within the fcc metastable matrix phase (Figure 7). This multiphase structure is characterized by high strength and ductility with the final properties being determined by both temperature and amount of deformation. Table 4 shows the effect of working temperature on the room temperature mechanical properties of the alloy. Additional strengthening results from an aging treatment in the range of 540 to 590°C (within the hcp + fcc two-phase field)

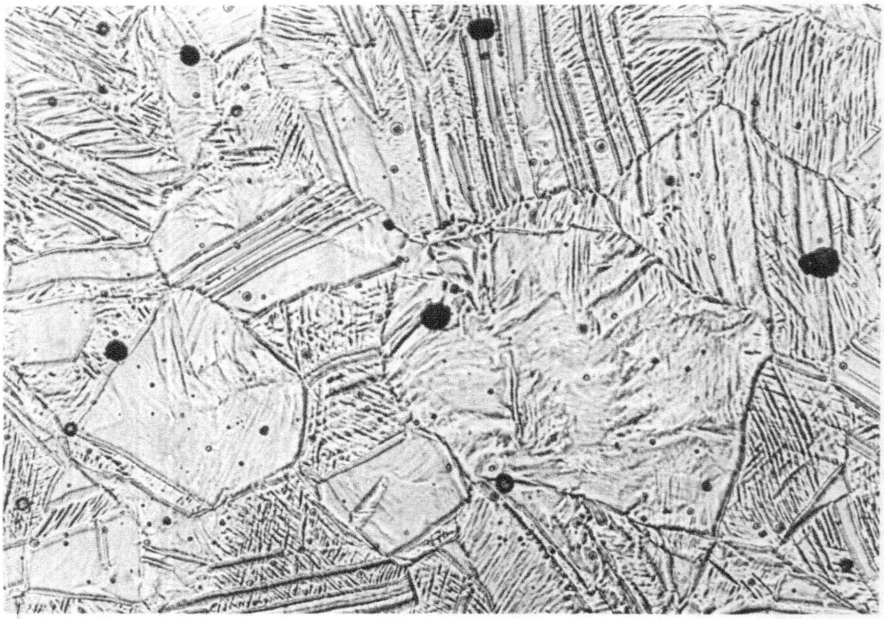

A

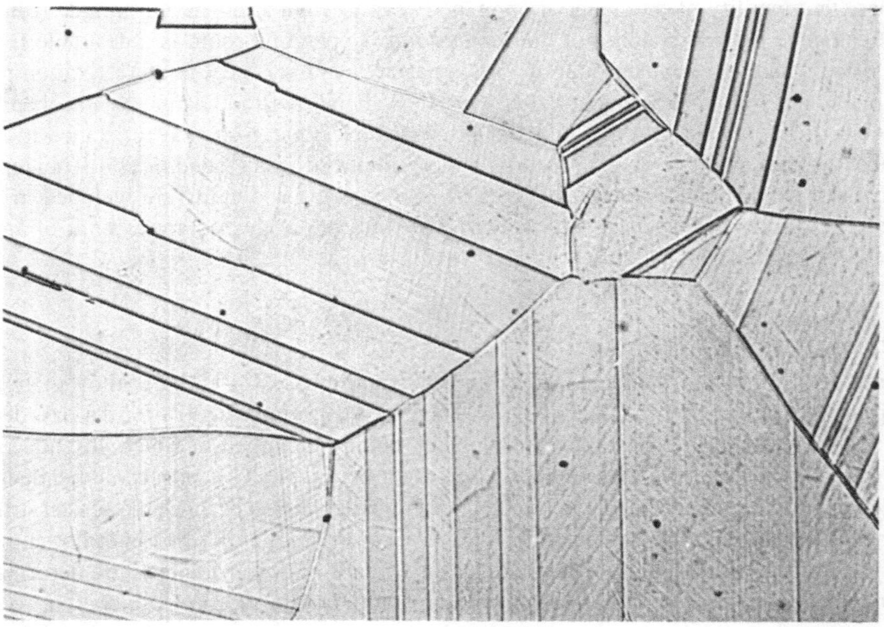

B

FIGURE 6. Microstructures of the wrought Co-Cr-W-Ni implant alloy. (A) Cold worked (magnification × 80); (B) annealed (magnification × 40).

for 4 hr followed by air cooling. Transformation of the strain-induced, hcp phase to the high temperature, stable fcc structure does not occur until annealing above 650°C. Thus, during the aging treatment, the hcp zones persist and provide sites for the precipitation of

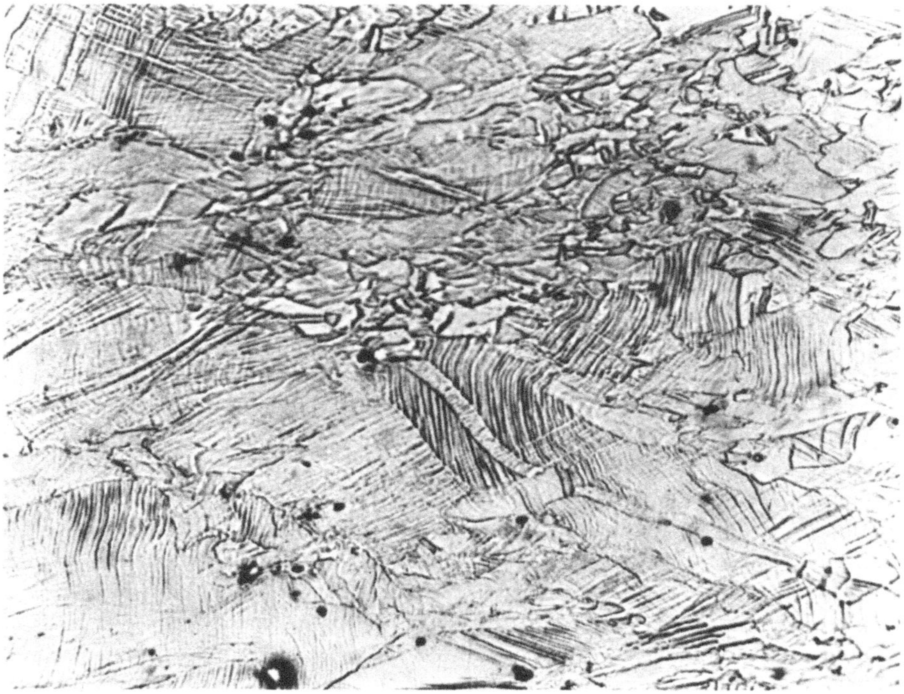

FIGURE 7. Microstructure of MP35N low temperature worked implant alloy; hcp bands are seen in the fcc matrix. (Magnification × 615.)

Table 4
MECHANICAL PROPERTIES OF MP35N

Treatment	σ_y(MPa)	σ_{ult}(MPa)	% Elongation	σ_{end} (10^7 c; R = −1)	Ref.
Annealed above 1050°C	300	800	40	340	26
Cold worked (50% red in area)	650	1000	20	435	26
Cold worked (55% red in area)	1413	1827	12	—	27
Cold worked (50% + aged)	—	—	—	405	26
Cold worked (55% + aged; 538°C/4 hr)	1999	2068	10	—	27
Hot forged (above 650°C)	—	1300	—	520	

a Co_3Mo intermetallic phase either within or at the interphase boundary of the hcp phase. This aging treatment results in further strength increase and a slight loss in ductility (Table 4).

Annealing at 1050°C for 1 to 2 hr results in loss in strength as indicated in Table 4. In this fully annealed condition, Co_3Mo precipitation does not occur on aging since the hcp zones, within which the precipitates form, do not exist.

Properties of MP35N, used for surgical implants, have been reported by a number of investigators. Lorenz et al.[26] reported on the significance of a 1% Ti addition to improve the forgeability of the alloy. Depending on the desired mechanical properties, typical products can be fabricated by:

1. Using a recrystallization (full) anneal at 1050°C for 1 hr to give a low yield strength (414 MPa), high-ductility (70% elongation) material

2. Hot forging above 650°C (in the fcc phase field), but below the recrystallization temperature (about 950°C), giving a medium hard to hard metal depending on the exact temperature and reduction used
3. Cold forging (below 650°C), where both strain hardening and the hcp phase formation influence strengthening
4. Cold forging followed by aging to give, as described above, the highest strengths (ultimate strength = 2550 MPa; yield strength = 2415 MPa; elongation = 4%)

Lorenz et al.[26] indicated that the hot forging process (i.e., forging above 650°C) gives better combinations of fatigue strength and ultimate strength compared to cold forging. For both hot and cold forged materials exhibiting an ultimate strength of 1300 MPa, an endurance limit (10^7) of 520 MPa (R = −1) was reported for the hot forged material compared to a 450 MPa endurance limit for a 13% cold-worked material. This difference was related to the poorer crack initiation resistance of the two-phase, cold-worked structure with its deformed hcp phase zones (identified as a martensite phase). These deformed hcp zones are susceptible to easier crack initiation.

Although the Co-Ni alloy is referred to as "multiphase", it is obvious that this feature is not unique to this particular Co-containing alloy. Also, single phase MP35N (fcc), as found in the fully annealed condition, can be used for implants. From the discussion of the cast and wrought Co-Cr-Mo alloys and the wrought Co-Cr-W-Ni alloys, fcc to hcp transformations occur under certain processing conditions for these alloys, and mechanical properties are not too unlike for the different alloys (see Tables 2, 3, and 4). With an appreciation of (1) the effect of processing on properties and (2) service load conditions, reliable implant design is possible with all these Co and Co-Ni base alloys and failure because of implant fracture should not occur.

C. 316L Stainless Steel

Most stainless steel implants used presently are wrought products of the composition shown in Table 1. Although some cast stainless steel components are made, the coarse grain size resulting in poor mechanical properties and the heterogeneous chemical composition resulting in poor corrosion resistance limits their use. Cast 316L stainless steel is not recommended for load-bearing implant applications.

Two ASTM standards have been specified for stainless steel for use in implant fabrication (Table 1). The lower-purity alloy (ASTM F55) is made by melting and casting ingots under air, inert gas, or vacuum atmosphere; the higher-purity type (ASTM F138 and F139) is produced by remelting a vacuum cast ingot either by vacuum arc remelting or by electro-slag remelting for impurity removal. For both types, two grades determined by carbon content are specified. The higher carbon stainless steel (C = 0.08% maximum) can be processed to higher strengths by working, but is susceptible to sensitization and intergranular corrosion. The lower carbon (0.03% maximum) alloy is less susceptible to this condition.

Most implants fabricated in Europe and North America at present are made from the cleaner vacuum arc or electro-slag remelt stock. The cast ingot is hot worked either by rolling, extrusion, or drawing to form bar or flat stock for further forming to implant shapes.

Final shaping of joint replacement prostheses is accomplished by a forging operation with typical forging temperatures being in the range of 950 to 1150°C. For the forging of 316 stainless steel in general, minimum periods at the high temperatures are recommended and for larger pieces a preheat at between 815 and 925°C is used thereby minimizing the holding time at higher temperatures.[28] The avoidance of long high-temperature soak times limits grain growth and excessive oxide scale formation. During forging, the dies are normally heated to 150 to 200°C in order to avoid chilling of the workpiece surface that can result in crack formation. Die lubrication with solutions such as colloidal graphite in kerosene or oil and graphite mixtures is also used. It is recommended that proper die lubrication be achieved prior to each forging blow to promote deformation and lessen the risk of surface cracking.

316L Stainless steel is fcc (austenitic) down to room temperature and, hence, alloy strengthening on working is accomplished through strain hardening only. Microstructures typical of 316L stainless steel implant alloys are shown in Figure 8.

Hughes and Jordan[29] reported on the mechanical properties of nitrogen annealed steels and indicated a modest increase in tensile and corrosion fatigue properties as a result of the higher nitrogen content of the steel (0.2 wt%). This added strengthening results from nitrogen atoms in solution.

Strengthening due to working will occur for deformations performed below the recrystallization and recovery temperature. This temperature will be dependent on the rate and type of straining imposed during the deformation process, but for 316L stainless steel will correspond to a temperature below about 500°C. 316 Stainless steel strain hardens fairly rapidly, and to prevent cracking on extensive cold working numerous intermediate annealing treatments are necessary. A 1050°C, 1/2-hr anneal will result in a fully softened steel with the properties as shown in Table 5. During hot working, the nature of the furnace atmosphere is important. It is essential that carburization of the steel does not occur so that susceptibility to sensitization is minimized. Also, oxidation should be minimized in order to avoid problems associated with oxide scale formation.

At temperatures in the range of 400 to 800°C, a Cr-rich, $M_{23}C_6$ phase is stable and will form at grain boundaries. The result of slow cooling or annealing in this range is the formation of this grain boundary phase with a concomitant loss of Cr from adjacent material. Since chromium is essential for passive film formation and corrosion resistance of this alloy, the consequence of a Cr-depleted region adjacent to grain boundaries is a zone particularly susceptible to corrosion. This results in the aforementioned "sensitized" structure and can lead to intergranular corrosion and cracking.

The use of low carbon material reduces but does not completely eliminate the possibility of sensitization. Thus, care must be taken constantly in the heat treatment of stainless steels. For this reason, intermediate annealing during working is carried out above 850°C; at this temperature the $M_{23}C_6$ phase is not stable and the Cr and C atoms remain in solution. The corrosion resistance of stainless steel components can be improved by use of a stabilizing treatment. This typically consists of an 885°C, 2-hr anneal followed by a stress relief anneal at 675°C.

As indicated in Table 5, cold-worked stainless steels are characterized by much higher strengths (static and fatigue).[30,31] A 20% reduction by cold working results in an increase in fatigue strength (10^7 c endurance limit) from 245 to 350 MPa.[2] Another result of cold work, however, is a reduction in corrosion resistance.[32,33]

The formation of an excessive oxide scale during intermediate annealing is undesirable since fragments of it can be introduced into the part on subsequent working and this can lead to regions prone to cracking. It is, therefore, essential that such scale is removed prior to further working operations. The removal of the scale can be accomplished by mechanical grinding or chemical dissolution.

The importance of grain size in determining fatigue strength for 316L stainless steel forgings was illustrated by Clemens et al.[34] A number of hip implants forged and annealed at 1050 and 1200°C were fatigue tested. The higher annealing temperatures (1200°C) resulted in regions of large grain size and a consequent loss in fatigue strength. This illustrates the importance of temperature control during processing of stainless steel implants.

Final trimming of forging flash from the shaped parts is achieved at 900 to 950°C. Implant cleaning subsequently is achieved mechanically by blasting using, typically, an alumina blasting media. The use of steel grit is not recommended unless an acid cleaning procedure

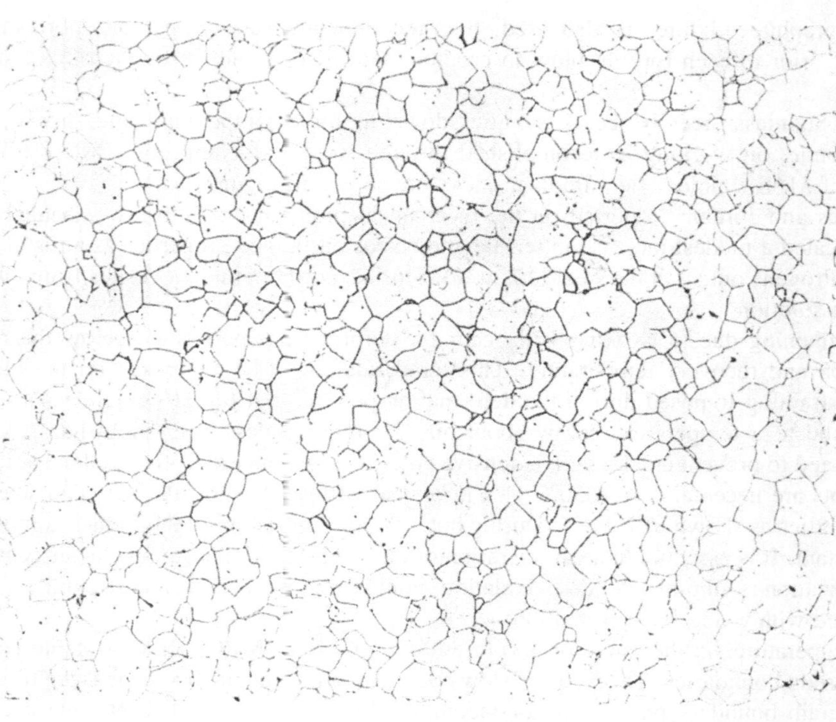

A

B

FIGURE 8. Microstructure of 316L stainless steel implants. (A) Hot forged hip implant; (B) Harrington rod indicating cold worked structure. (Magnification × 100.)

Table 5
MECHANICAL PROPERTIES OF STAINLESS STEEL FOR SURGICAL IMPLANTS

Treatment	σ_y(MPa)	σ_{ult}(MPa)	% Elongation	σ_{end} (10^7 c; R = −1)	Ref.
Annealed; 1050°C/1/2 hr	211	645	68	190—230	30
Annealed; 1050°C/1/2 hr	207	517	40	—	31
Cold worked	1160	1256	6	530—700	30
Cold worked	689	862	12	—	31
Cold worked (20% red in area)	—	—	—	345	30
Annealed + N_2	380	700	46	269	30

follows because of the possible surface contamination of the implant with steel, resulting in galvanic corrosion cells in vivo.

After cleaning, all stainless steel parts must be passivated using one of a number of specified procedures.[35] This process assures the formation of a dense, coherent oxide film on the implant surface that renders the stainless steel usable for implant applications. Often, one-off stainless steel parts are machined for special applications by individual researchers or hospital workers. It is important that such parts also be passivated prior to their in vivo use.

D. Ti and Ti6Al4V Alloys

Titanium is a reactive element and as such requires special care during its processing. Both pure titanium and its alloys will react readily with common gases such as oxygen, hydrogen, and nitrogen, resulting in changes in mechanical properties. Excessively high gas contents would result in limited ductility. To form as high a purity material as possible, double vacuum (or inert gas), consumable electrode arc melting methods are used.[36] By this method, trace element contamination is kept to a minimum (Table 1). For the Ti6Al4V alloy, a columnar grain structure with solute-enriched interdendritic regions results in the as-solidified ingot. Prior to any extensive deformation of the alloy, this structure must be homogenized.

Unalloyed titanium (C.P. Ti, see Table 1) is close packed hexagonal (α-phase) at temperatures below 882°C, while above this temperature a body centered cubic β-phase is stable. Typical forging temperatures for unalloyed titanium are about 910°C (in the β-phase field). The unalloyed form of titanium has found use as an implant alloy in the past, but its limited mechanical properties, although making fabrication easier, renders it unsuitable for highly loaded implants. The properties of the unalloyed titanium (like those of the Ti6Al4V alloy) are affected strongly by oxygen content.[37] Increased levels of interstitial oxygen result in increased strengths and reduced ductilities (Table 6).

The Ti6Al4V alloy (see Table 2 for chemical composition) is characterized by a two-phase, α- + β-phase field below about 990°C. Shaping is achieved by hot working with recommended forging temperatures being 50 to 150°C below the α + β- to β-transition temperature. Working of the material can be continued down to about 760°C, but below this temperature there is a danger of surface cracking. Hot forging also can be achieved by heating in the β-phase field (>990°C). At least 50% of the reduction during forging, however, should occur after the part has cooled to below the transition temperature in order to achieve desirable microstructures.[36] This structure ideally should consist of about 15 to 30% by volume of primary α-phase within a matrix of transformed β (Figure 9). The transformation product will consist of an α-prime phase (a martensitic transformation product) and α-phase. The avoidance of long lath-like α/β boundaries is recommended in order to attain reasonable fatigue properties.[38]

Table 6
MECHANICAL PROPERTIES OF Ti AND Ti6Al4V SURGICAL IMPLANT ALLOYS

Treatment	σ_y(MPa)	σ_{ult} (MPa)	% Elongation	σ_{end} (10^7 c; R = −1)	Ref.
C.P. titanium,[a] annealed	385	530	23	—	51
Ti6Al4V, β-annealed (1030°C); furnace cooled to 800°C; air cooled	838	948	12.5	440	37
α-β worked (650 to 700°C), annealed at 700°C	1036	1147	12.5	670	37
Mill annealed	966	1000	43(%R.A.)[b]	—	43

[a] C.P. = commercial purity.
[b] R.A. = reduction in area.

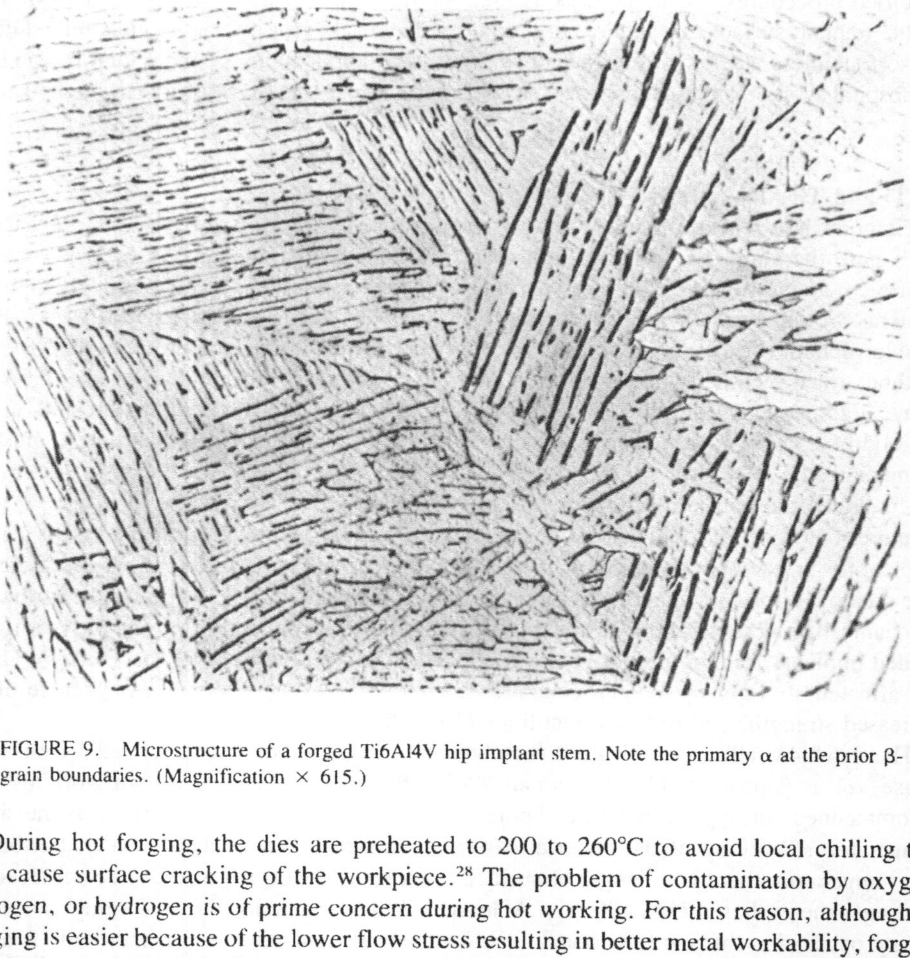

FIGURE 9. Microstructure of a forged Ti6Al4V hip implant stem. Note the primary α at the prior β-grain boundaries. (Magnification × 615.)

During hot forging, the dies are preheated to 200 to 260°C to avoid local chilling that can cause surface cracking of the workpiece.[28] The problem of contamination by oxygen, nitrogen, or hydrogen is of prime concern during hot working. For this reason, although β-forging is easier because of the lower flow stress resulting in better metal workability, forging within the α- + β-phase field results in less contamination because of the lower gas diffusion rates. To further decrease gas diffusion into forged parts, billets to be hot forged often are coated with a glass-type coating material (soda barium or borosilicate glass).[35] This glassy coating can be deformed at 680°C or so. As well as serving as a protective coating, it also acts as a lubricant to facilitate forging. Other lubricants that are used are colloidal graphite in water, graphite and molybdenum disulfide suspensions in water or oil bases, metal oxides, graphites, and silicone base materials.

To avoid excessive hydrogen contamination which can lead to the formation of embrittling needle-like hydride precipitates throughout the metal matrix, mildly oxidizing atmospheres are used during preheating of billets to be forged.[35] As a result, oxygen enrichment at the surface resulting in a region of α-phase (the α-case) occurs, (O_2- and N_2-like Al and C are α-stabilizers). This α-case is undesirable because of its poorer mechanical properties, particularly ease of cracking.

The removal of surface oxide scale and the α-case is accomplished mechanically either by grinding or by grit blasting using zirconia grit (the use of ZrO_2 is necessary since the reactive nature of titanium would result in its reduction of other less stable particles). Chemical cleaning using a nitric-hydrofluoric acid pickling solution, 40% HNO_3 + 1% HF + H_2O, at 20 to 40°C, also is used for surface preparation.

A number of studies have been reported on the effect of processing of Ti6Al4V on fatigue properties.[38-49] These studies have mostly been stimulated by the aerospace applications of this alloy. The results suggest that fatigue strengths are strongly related to the α-grain size and morphology. The highest fatigue strengths correspond to fine equiaxed α-grains with spheroidal β-grains.[38] Such a structure was achieved by Stubbington and Bowen by annealing bar stock in the β-phase field (at 1030°C), furnace cooling to 800°C, and then air cooling to room temperature. Bar stock so treated then was worked in the α- + β-phase field (at 650 to 700°C). After working, the samples were annealed at 700°C. The fatigue properties of samples so treated were dependent on the percent reduction as indicated in Table 6.

The appearance of tissue discoloration adjacent to Ti alloy implants has caused concern because of possible excess metal ion release and chronic tissue reaction. Solar et al.[50] have suggested that this discoloration is related to needle-shaped oxide microspicules that are abraded from the implant surface. A proper passivation treatment such that an irregular surface oxide morphology does not occur was recommended in order to prevent this phenomenon. This illustrates the possible importance of passivation treatment and solutions used in this treatment in forming a suitably coherent, dense and regular oxide layer on surgical implants.

E. Combined Metal Systems

1. Welded Structures

The effect of processing on metal properties has been illustrated in the discussion of cast and wrought alloys and the improved fatigue strengths of wrought alloys compared to cast alloys have been noted. However, wear resistance for metal implant alloys is greatest for the cast Co-Cr-Mo alloy,[51] while the use of Ti6Al4 alloy as a bearing surface remains questionable.[52] In order to achieve a combination of high wear resistance of the bearing surface and high fatigue strength of the stem region of a femoral hip implant, composite structures consisting of cast Co-Cr-Mo femoral head components welded to MP35N[53] or Ti6Al4V[54] stem components have been fabricated. The two dissimilar metal components are joined by electron beam or tungsten inert gas (TIG) welding after which the bearing surface is machined to produce the required smooth surface finish. The heat-affected weld zone is confined to a region remote from the critical highly stressed regions of the stem so that the annealing effect is not detrimental to overall implant strength. In choosing metals that form stable passive oxide films, low corrosion rates result despite the dissimilar metal combinations[55] so that useable combined metal implants are formed.

2. Porous-Surfaced Implants

In order to achieve implant fixation by tissue ingrowth a number of different porous-surface structured implant designs have been developed, and some of these are presently being evaluated clinically.[56-58] By confining the porous structure to the surface of the implant, overall mechanical strength can be retained through the use of a solid metal core. Two general approaches to forming these porous-surface structures have been developed, namely, the sintering of metal particles (powders[59] or fibers[60]) to the solid substrate surface or specialized casting processes (void metal composite).[61]

The powder metal-made porous-surfaced implants are fabricated by placing atomized alloy powders on the surface of a solid component and annealing at elevated temperatures (1200 to 1300°C) in a reducing atmosphere. By forming a mixture of metal particles in an organic binder, complex shapes of uniform coating thickness can be formed with relative ease. By controlling the sintering conditions, desired strengths and porosities of the surface layer can be achieved. The exact sintering conditions are determined by the alloy powder used and its size distribution with coarser powders requiring higher temperatures and/or longer sintering times to attain desired strengths.

The fiber-coated systems that have been reported have been formed from stainless steel or titanium fibers.[62] The use of kinked fibers provides a mechanically more compliant surface and it has been claimed that this is beneficial to bone remodeling at the interface.[62] Again, a high-temperature sintering operation is necessary for bonding of the fibers to each other and to the solid substrate. Sintering conditions depend on fiber type and size. For 0.19-mm diameter, 6.35-mm length titanium fibers, a 1093°C, 2-hr sinter in a vacuum better than 10^{-4} mmHg was reported.[60] For the fiber sintered systems, initial fiber placement was achieved by pressing the fibers onto a surface using an uniaxial compacting press. Ducheyne et al.[63] reported cold isostatic pressing of 316L stainless steel fibers (50- and 100-μm diameter and 4-mm length) on the surface of a solid metal substrate followed by 1250°C, 2- or 4-hr sinter in a hydrogen atmosphere. The difficulties of specimen containment during cold isostatic compaction were discussed. The use of isostatic pressing did offer the advantage over uniaxial pressing of porous coating complex shapes with undercuts and curved surfaces. With uniaxial pressing, undercut zones cannot be treated without very complex die designs and uniform pressuring of curved surfaces is difficult.

Void metal composites (VMC) are prepared by mixing low melting point, sacrificial magnesium or calcium particles with titanium or stainless steel powders.[61] The powder mixture is vibrated at high frequency to obtain good powder packing. It then is placed in a vacuum containment vessel and compacted using high energy rate forming at 530°C. Subsequently the containment material is removed and the Mg or Ca particles are vaporized using a high-temperature (1000°C) vacuum anneal. The resulting porous material is given a final sinter at 1400°C or so after which a passive oxide film is formed on the porous part by electrolytic anodization. By this process, close control of pore size can be achieved through the choice of sacrificial powder size. However, the process is complex, and to the best of the author's knowledge, no surgical implants for clinical use are being formed by this method.

A "porous-surfaced" Co-Cr-Mo implant introduced by Judet[63] provides surface irregularities 200 μm to 1 mm large and deep for bone ingrowth. The "porous" structure unlike that for the VMC implants or for the fiber or powder sintered systems appears to be a surface irregularity only. True, three-dimensional, interconnected porosity is not formed. Although details of manufacture could not be found in the literature, such structures could be formed by the use of investments with rough inner surfaces.

Porous-surfaced metal powder-made implants have been developed by the author and co-workers.[59] Although any of the implant alloys could be used, it is suggested that this method be limited to the more passive alloys only. Thus, the use of stainless steel is not recommended for porous-surfaced implant design.

The method of fabrication allows the uniform coating of complex shapes since the solid implants are coated by dipping or spraying with a metal powder-organic binder slurry followed by the sintering operation. For the cast Co-Cr-Mo alloys, an inert gas atomized powder is used. A powder size range of 50 to 400 μm has been found suitable for bone ingrowth,[65] and by proper sintering treatments very strong and fatigue resistant surface coatings can be formed.[59] Despite the larger surface area of these types of implants, they appear biocompatible.[66]

Concern over the possible effects of the elevated temperature sintering treatment on the structure and properties of the substrate has prompted fatigue studies of porous surfaced Co-Cr-Mo[11] and Ti6Al4V[67] alloys. The results indicate that mechanical properties can be retained provided suitable heat treatments are used. However, improper heat treatments can cause deterioration of mechanical properties.[11,67]

A thorough understanding of processing on structures and properties of these alloys is particularly important in forming these newer implant designs. It is appropriate that the conclusion of this chapter in dealing with the newer concepts of metal implants for orthopedic applications should illustrate this important process-properties-performance relationship. It emphasizes the need for critical analysis of any new development for "better" implants.

ACKNOWLEDGMENTS

The author would like to express his thanks to Messrs. T. Kilner (University of Toronto, Department of Metallurgy and Materials Science) and A. Lippincott (Depuy) for provision of samples and photographs. The assistance of F. Neub and B. Stopay with photographs and typing is acknowledged as are the critical comments of Dr. G. C. Weatherly and Dr. W. J. Bratina (University of Toronto, Department of Metallurgy and Materials Science).

REFERENCES

1. American Society for Metals, *Metals Handbook,* Vol. 5, ASM, Metals Park, Ohio, 1970, 237.
2. **Tottle, C. R. and Wright, J. C., Eds.,** in *Inst. of Metallurgists: Modern Metallurgical Texts,* No. 5 *Mechanical Treatment of Metals,* Elsevier, New York, 1968.
3. **Ladanyi, T. J., Meyers, G. A., Pilliar, R. M., and Weatherly, G. C.,** Fracture toughness of powder forged Cr-Mn alloy steels, *Metal Trans. A,* 6a, 2037, 1975.
4. **Pilliar, R. M., Hollingbery, D. R., and Fossen, W. F.,** The integrity of p/m Ni-Mo steels determined by fracture toughness, *Int. J. Powder Metall. Powder Technol.,* 13(2), April 1977.
5. **Williams, D. F. and Roaf, R.,** in *Implants in Surgery,* W. B. Saunders, Philadelphia, 1973.
6. **Kilner, T., Pilliar, R. M., Weatherly, G. C., and Allibert, C.,** Phase identification and incipient melting in a cast Co-Cr surgical implant alloy, *J. Biomed. Mater. Res.,* 16, 63, 1982.
7. **Hollander, R. and Wulff, J.,** New technology for mechanical property improvement of cast Co-Cr-Mo-C surgical implants, *Eng. Med.,* 3(4), 8, 1974.
8. **Hodge, F. G. and Lee, T. S., III,** Effect of processing on performance of cast prosthetic alloys, *Corrosion,* 31(3), 111, 1975.
9. **Hollander, R. and Wulff, J.,** Improved processing of Co-Cr-Mo-C surgical implant alloys, *M. Eng. Q.,* 14(4), 37, 1974.
10. **Clemow A. J. T. and Daniell, B. L.,** Solution treatment behaviour of Co-Cr-Mo alloy, *J. Biomed. Mater. Res.,* 13, 265, 1979.
11. **Elve, A., Kilner, T., Pilliar, R. M., and Weatherly, G. C.,** Structure property relationships for porous coated Co-Cr surgical implants, paper presented at 7th Annu. Soc. for Biomaterials Meeting, Troy, N.Y., May 1981.
12. **Galante, J. O., Rostoker, W., and Doyle, J. M.,** Failed femoral stems in total hip prostheses. A report of six cases, *J. Bone Jt. Surg.,* 57a, 230, 1975.
13. **Weeton, J. W. and Signorelli, R. A.,** Effect of heat treatment upon microstructures, microconstituents, and hardness of a wrought cobalt base alloy, *Trans. Am. Soc. Met.,* 47, 815, 1955.

14. **Müller, M.,** Endocast — a new material for long-term implants, paper presented at Osteo Symposium, Grindelwald, West Germany, June 24, 1977. Osteo A. G. CH-2545 Selzack.
15. **Cohen, J., Rose, R. M., and Wulff, J.,** Recommended heat treatment and alloy additions for cast Co-Cr surgical implants, *J. Biomed. Mater. Res.*, 12, 935, 1978.
16. **Weatherly, G. C.,** A determination of the punching stress at the interface of particles during rapid quenching, *Met. Sci. J.*, 2, 237, 1968.
17. **Devine, T. M., Kummer, F. J., and Wulff, J.,** Wrought Co-Cr surgical implant alloys, *J. Mater. Sci.*, 7, 126, 1972.
18. **Devine, T. M. and Wulff, J.,** Cast versus wrought Co-Cr surgical implant alloys, *J. Biomed. Mater. Res.*, 9, 151, 1975.
19. **Vander Sande, J. B., Coke, J. R., and Wulff, J.,** A transmission electron microscopy study of the mechanisms of strengthening in heat-treated Co-Cr-Mo-C alloys, *Metall. Trans. A*, 7a, 389, 1976.
20. Vitallium (FHS) Technical Monograph: Howmedica Inc., Rutherford, N.J., 1979.
21. **Bardos, D. I.,** High strength Co-Cr-Mo alloy by hot isostatic pressing of powder, *Biomater. Med. Devices Artif. Organs*, 7(1), 73, 1979.
22. Zimmer Materials Research Group, Zimaloy, Micro-grain, Tech. Rep., Warsaw, Ind., January 24, 1978.
23. **Hughes, A. N. and Lane, R. A.,** The development of high tensile properties in a wrought Co-Cr-Ni-W alloy of potential orthopaedic surgery use, *Eng. Med.*, 5(2), 38, 1976.
24. Medical Metals, Richards Publ. No. 3922, Richards Manufacturing Co., Memphis, Tenn., 1980.
25. **Straumann, R.,** *About Wrought Co-Ni-Cr-Mo-W-Fe Alloy (Syntacoben) for Surgical Implants,* (Documentation to ASTM) American Society for Testing and Materials, Philadelphia, 1976.
26. **Lorenz, M., Semlitsch, M., Panic, B., Weber, H., and Wilhert, H. G.,** Fatigue strength of cobalt-base alloys with high corrosion resistance for artificial hip joints, *Eng. Med.*, 7(4), 241, 1978.
27. Multiphase MP35N Alloy, Tech. Data Latrobe Steel Co., Latrobe, Pa., 1980.
28. American Society for Metals, Metals Handbook, Vol. 5, 8th ed., ASM, Metals Park, Ohio, 1970, 113.
29. **Hughes, A. N. and Jordan, B. A.,** Corrosion fatigue properties of treated 316S12 stainless steel, *Eng. Med.*, 1, 117, 1972.
30. **Smith, D. J. E. and Hughes, A. N.,** The influence of frequency and cold work on the fatigue strength of 316 stainless steel in air and 0.17 *M* saline, AWRE Rep. 44/83/189, Atomic Weapons Research Establishment, Aldermaston, Berkshire, U. R., April 1977.
31. **Fraker, A. C. and Ruff, A. W.,** Metallic surgical implants: state of the art, *J. Met.*, 29(5), 22, 1977.
32. **Sutow, E. J., Pollack, S. R., and Korostoff, E.,** An in vitro investigation of the anodic polarization and capacitance behaviour of 316L stainless steel, *J. Biomed. Mater. Res.*, 10, 671, 1976.
33. **Cigada, A., Mazza, B., Pedeferri, P., and Sinigaglia, D.,** Influence of cold plastic deformation on critical pitting potential of AISI 316L and 304L steels in an artificial physiological solution simulating the aggressiveness of the human body, *J. Biomed. Mater. Res.*, 11, 503, 1977.
34. **Clemens, D., Galante, J., and Rostoker, W.,** The influence of grain size on the fatigue behaviour of annealed 316LVM stainless steel , *J. Biomed. Mater. Res.*, 13, 437, 1979.
35. ASTM Standard F86-76, *Surface Preparation and Marking of Metallic Surgical Implants,* American Society for Testing and Materials, Philadelphia, 1976.
36. **Brookes, R.,** The forging characteristics of titanium and its alloys, in *Material Yield and Improved Technology*, British Iron and Steel Research Association, 1971, chap. 5.
37. **Williams, D. F.,** Titanium as a metal for implantation. I. Physical properties, *J. Med. Eng. Technol.*, p. 195, July 1977.
38. **Stubbington, C. A. and Bowen, A. W.,** Improvements in the fatigue strength of Ti-6Al-4V through microstructure control, *J. Mater. Sci.*, 9, 941, 1974.
39. **Hirth, J. P. and Froes, F. H.,** Interrelations between fracture toughness and other mechanical properties in titanium alloys, *Metall. Trans. A*, 8a, 1165, 1977.
40. **Kuhlman, G. W. and Billman, R. R.,** Selecting processing options for high-fracture toughness titanium airframe forgings, *Metal Progress*, p. 39, March 1977.
41. **Yoder, G. R., Cooley, L. A., and Crooker, T. W.,** Quantitative analysis of microstructural effects on fatigue crack growth in Widmanstatten Ti6Al4V and Ti 8Al 1Mo 1V, *Eng. Fract. Mech.*, 11, 805, 1979.
42. **Ryder, J. T., Krupp, W. E., Pettit, D. E., and Hoeppner, D. W.,** Corrosion-fatigue properties of recrystallization annealed Ti-6Al-4V, *Corrosion-Fatigue Technology,* ASTM STP 642, American Society for Testing and Materials, Philadelphia, 1978, p. 202.
43. **Benson, D. K., Grosskreutz, J. C., and Shaw, G. C.,** Mechanisms of fatigue in mill-annealed Ti-6Al-4V at room temperature and 600°F, *Metall. Trans.*, 3, 1239, 1972.
44. **Sherman, R. G. and Kessler, H. D.,** Investigation of the heat treatability of the 6% aluminum- 4% vanadium titanium-base alloy, *Trans. Am. Soc. Met.*, 48, 657, 1956..
45. **Lasalmonie, A. and Loubradou, M.,** The age hardening effect in Ti-6Al-4V due to α precipitation in the β grains, *J. Mater. Sci.*, 14, 2589, 1979.

46. **Yoder, G. R., Cooley, L. A., and Crooker, T. W.,** Fatigue crack propagation resistance of beta annealed Ti-6Al-4V alloys of differing interstitial oxygen contents, *Metall. Trans. A*, 9a, 1413, 1978.
47. **Bania, P. J. and Eylon, D.,** Fatigue crack propagation of titanium alloys under dwell-time conditions, *Metall. Trans. A*, 9a, 847, 1978.
48. **Eylon, D. and Pierce, C. M.,** Effect of microstructure on notch fatigue properties of Ti-6Al-4V, *Metall. Trans. A*, 7a, 111, 1976.
49. **Smith, C. J. E. and Hughes, A. N.,** The corrosion fatigue behaviour of a titanium-6 w/o aluminium-4 w/o vanadium alloy, *Eng. Med.*, 7(3), 158, 1978.
50. **Solar, R. J., Pollack, S. R., and Korostoff, E.,** Titanium release from implants: a proposed mechanism, in *Corrosion and Degradation of Implant Materials*, ASTM STP684, American Society for Testing and Materials, Philadelphia, 1979, p. 161.
51. **Mears, D. C.,** Metals in medicine, *Intl. Met. Rev.*, p. 119, June 1977.
52. **McKellop, H., Clarke, I., Markolf, K. and Amstutz, H.,** Wear properties of new high strength alloys for prosthetic joints, in Trans. 25th Annu. Meeting, Orthopaedic Research Society, San Francisco, Calif., Vol. 4, 1979, 71.
53. **Sury, P. and Semlitsch, M.,** Corrosion behavior of cast and forged cobalt-based alloys for double-alloy joint endoprostheses, *J. Biomed. Mater. Res.*, 12, 723, 1978.
54. **Russin, L. A.,** The Sivash Total Hip Prosthesis, U.S. Surgical Corp., New York, 1974.
55. **Mears, D. C.,** The use of dissimilar metals in surgery, *J. Biomed. Mater. Res.*, 6, 133, 1975.
56. **Welsh, R. P., Pilliar, R. M., and Macnab, I.,** Surgical implants — the role of surface porosity in fixation, *J. Bone Jt. Surg.*, 53a(5), 963, 1971.
57. **Homsy, C. A., Cain, T. E., Anderson, M. S., and King, J. W.,** Dynamic stabilization of implanted prostheses, Abst. ORS Proc., *J. Bone Jt. Surg.*, 52a, 604, 1970.
58. **Lambert, E., Galante, J., and Rostoker, W.,** Fixation of skeletal replacement by fiber metal composites, *Clin. Orthop. Rel. Res.*, No. 87, 303, 1972.
59. **Pilliar, R. M., Cameron, H. U., and Macnab, I.,** Porous surface layered prosthetic devices, *Biomed. Eng.*, 10(4), 126, 1975.
60. **Galante, J., Rostoker, W., Lueck, R., and Ray, R. D.,** Sintered fiber metal composites as a basis for attachment of implants to bone, *J. Bone Jt. Surg.*, 53a(1), 101, 1971.
61. **Nilles, J. L., Karagianes, M. T., and Wheeler, K. R.,** Porous titanium alloy for fixation of knee prostheses, *J. Biomed. Mater. Res. Symp.*, 5 (Part 2319), 328, 1974.
62. **Rostoker, W., Galante, J. O., and Shen, G.,** Some mechanical properties of sintered fiber metal composites, *J. Test. Eval.*, 2(2), 107, 1974.
63. **Ducheyne, P., DeMeester, P., and Aernoudt, E.,** Isostatically compacted metal fiber porous coatings for bone ingrowth, *Powder Metall. Int.*, 11, 115, 1979.
64. **Judet, R.,** Prothèse totale en poro-métal habitable, Ets. R. Tornier — 38330, Saint Ismier, France.
65. **Bobyn, J. D., Pilliar, R. M., Cameron, H. U., and Weatherly, G. C.,** The optimum pore size for the fixation of porous-surfaced metal implants by bone ingrowth, *Clin. Orthop. Rel. Res.*, 150, 263, 1980.
66. **Margolian, S.,** Differential Ion Diffusion of Porous Coated and Uncoated Metal Implants in Soft Tissue and Its Biological Implications, M. Eng. thesis, University of Toronto, Ontario, 1980.
67. **Mooz, A.,** Mechanical Properties of a Surgical Grade Titanium Alloy, M.A. Sc. thesis, University of Toronto, Ontario, 1980.

Chapter 5

MANUFACTURING PROCESSES OF CERAMICS

Robert H. Doremus

TABLE OF CONTENTS

I. INTRODUCTION

Ceramics have many desirable properties as biomaterials. They are nontoxic and inert to most biological fluids, use inexpensive and readily available raw materials, and can be made in a wide variety of sizes and shapes. Their main disadvantage is brittleness; however, in many applications this property is not a serious problem. The use of hydroxyapatite as a bone implant and tooth replacement is especially attractive because it is chemically identical to the mineral in bones and teeth and is readily accepted as an implant.

In this review raw materials for ceramics are described, and then the making and forming of the different types of ceramics, crystalline ceramics, glasses, glass ceramics, and carbons, are considered separately because the processing of each is distinctive.

There is a good account of the processing of crystalline ceramics in the first edition of Kingery's book;[1] more details are given by Norton.[2] There is much information on the structure and properties of ceramics in Kingery et al.[3] Douglass and Frank[4] have described the history and technology of glassmaking; Tooley[5] has collected much information on glass manufacture. The structure and properties of glass are summarized by Doremus,[6] and making and properties of glass ceramics described by McMillan.[7] The manufacture of solid carbons is reviewed in the *Encyclopedia of Chemical Technology*.[8]

II. RAW MATERIALS

A. Clay

Clay is the traditional raw material for crystalline ceramics and is still used today for a wide variety of materials, such as porcelain, dinner ware, tile, and brick. The clay minerals are hydrated aluminosilicates; chemical formulas for some clays are kaolinite, $Al_2O_3{\cdot}2SiO_2{\cdot}2H_2O$; montmorillonite, $5/6\ Al_2O_3{\cdot}1/6\ Na_2O{\cdot}1/3\ MgO{\cdot}4SiO_2{\cdot}H_2O$; mica, $3Al_2O_3{\cdot}K_2O{\cdot}6SiO_2{\cdot}2H_2O$. In nature, clays are always contaminated with a variety of other minerals, such as feldspar, quartz, alkali, and alkaline earth salts, and with organic matter. Different ions, such as iron, calcium, magnesium, sodium, and potassium, can substitute for ions of the same charge in the clay structure, giving a wide variety of chemical compositions of clays. Clays are often used directly as mined, and their properties and reactions in production depend strongly on impurities. Clays can be purified by pretreatment, especially to remove alkalis and alkaline earths to give a higher melting and more uniform product.

Clay is valuable because it is "plastic" and easily formed when wet. It also partially fuses over a range of temperatures during firing, depending upon its composition; this flux aids in forming a strong and dense body from a mixture of raw materials.

Although clay seems to be amorphous, it is actually made up of fine plate-like crystals from about 0.1 to 10 μm in size. This shape helps to give clay its forming properties, because when mixed with water these plates slide over one another easily.

B. Other Silicates

Quartz (SiO_2) is available as sand or chunks. It can be one of the purest natural materials and is a major constituent of glasses, porcelain, glazes, and enamels. Quartz can also be made by "hydrothermal" precipitation from water under pressure above its normal boiling temperature.

Feldspars are anhydrous aluminosilicates that contain alkali or alkaline earth ions, mainly sodium, potassium, or calcium. Typical minerals are orthoclase, $K_2O{\cdot}Al_2O_3{\cdot}6SiO_2$; albite, $Na_2O{\cdot}Al_2O_3{\cdot}6SiO_2$; and anorthite, $CaO{\cdot}Al_2O_3{\cdot}2SiO_2$. Feldspars are usually mixed with other minerals, so they are often separated from them before being used for pottery, porcelain, enamels, and glass. Talc is a magnesium silicate $3MgO{\cdot}4SiO_2{\cdot}S_2O$ that is available relatively pure and is used in making tile and highly resistive electrical components.

C. Other Oxides

Aluminum oxide is made from the mineral bauxite, which is hydrated Al_2O_3. Alumina is very hard and refractory and is used in furnaces and crucibles, in electronic components, and is being considered as a tooth replacement material and in artificial joints.[9]

Apatite is the main phosphate-containing mineral, and as hydroxyapatite, $Ca_{10}(PO_4)_6 \cdot H_2O$, is the mineral in bones and teeth. Synthetic hydroxyapatite is being tested as a bone and tooth replacement material.[10] The commercial bulk carbonates of sodium (soda ash) and calcium are used for making glass and as fluxes in other ceramics.

D. Carbons

Pitch is the usual raw material for making industrial solid carbons. It is the residue from distillation of petroleum or other organic material such as coal or wood and is composed mainly of aromatic hydrocarbons of average molecular weight of 300 daltons. It is hard at room temperature, but softens in the temperature range of 50 to 180°C. Coke is also a raw material for industrial carbons and can be made by heating pitch. It is solid up to high temperatures and has an average molecular weight of greater than 3000 daltons.

III. MANUFACTURING OF CRYSTALLINE CERAMICS

The large majority of inorganic crystalline ceramics are made in three steps: forming of a powder to the desired shape, partial drying, and firing. The size, shape, and chemical composition of the crystalline powders strongly influence subsequent processing, so powder preparation is considered first. Then methods of forming and drying are considered and some methods of making particular ceramics of biomedical interest are outlined. Finally a variety of special methods of making crystalline ceramics are described. The mechanisms of sintering a powder to a dense object are treated in the following section because of their distinctive importance.

A. Powder Processing

In traditional ceramics, powdered raw materials such as clay, feldspar, and sand are used directly as mined without any pretreatment. However, control of crystalline particle size and size distribution of the clay and of impurities can improve properties and simplify processing. For materials such as alumina, electronic ceramics, and most biomaterials, mechanical, electrical, chemical, and optical properties are strongly influenced by purity and particle size and distribution of the starting material.

Particle size and distribution of the starting materials can be controlled by crushing, milling, and classifying. Classifying is achieved by sieving or by air or water separation. Impurities can be removed by chemical treatment, usually by various aqueous solutions such as acids. Flotation and filtration may be involved.

Calcining involves heating a material to an intermediate temperature below that of firing, e.g., about 800°C. This treatment can partially decompose starting materials and remove undesirable impurities by decomposition (water, CO_2) or by chemical reaction (oxidation of organic matter).

For special ceramics that must be pure and have controlled particle size and size distribution, precipitation from solution is a desirable method of preparing starting powders. An example is the precipitation of hydroxyapatite from basic solution as a starting material for a dense ceramic for bone implants.[10] Solutions of calcium nitrate and ammonium phosphate were mixed and ammonium hydroxide was added to a pH of 11 to 12. The precipitate was centrifuged and filtered and then partially dried and fired directly. An electron micrograph of the precipitated crystals is shown in Figure 1. Other chemical methods such as hydrothermal and gel techniques can be used to produce pure, controlled starting powders.

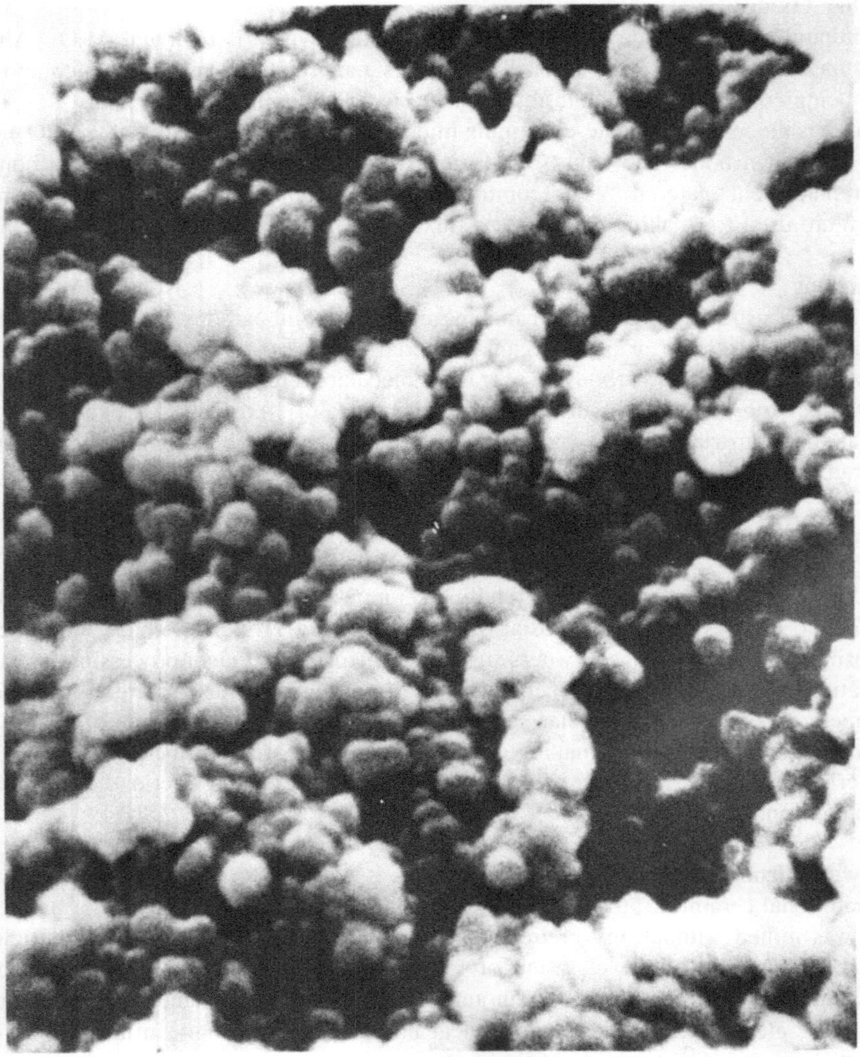

FIGURE 1. Hydroxyapatite crystals precipitated from solution; average diameter ~50 nm. (Magnification × 75,000.)

B. Forming

Powders are usually pressed damp in metal dies at a high enough pressure to form material strong enough to hold together until it is fired. For pure powders an organic binder such as polyvinyl alcohol helps to bind the powder together. In isostatic or hydrostatic molding, the powder is placed in a rubber die or sheath and pressure is applied to it through a fluid. Cylindrically symmetrical shapes can be extruded through a die to give the desired shape.

Slip casting is used for forming complicated shapes from a slip or suspension of powder particles in a slurry. The slip is poured into a porous mold made of plaster of Paris, and the mold absorbs water from the slip, leaving a hard coating inside the mold. This method can be used either for solid or hollow shapes.

C. Drying

A drying step is critical in preparing the formed pieces for firing. If too much water or binder is left in the material, they will be given off as the temperature is raised and break apart the ware. If the piece is too dry, it will crumble before it can sinter to a dense body. Drying is usually carried out at about 100°C or lower. It may also be necessary to control the humidity to regulate the rate of drying. Binder is removed by heating to a higher temperature in air to oxidize the organic binder material and drive off the resultant water and carbon dioxide.

D. Porcelain

Porcelains contain various mixtures of clay, potash feldspar, $K_2O{\cdot}Al_2O_3{\cdot}6SiO_2$, and quartz sand. A typical hard porcelain for whiteware contains 50% clay, 25% feldspar, and 25% sand. The feldspar provides alkali that melts at a relatively low temperature and binds the remaining constituents together. During firing the clay dehydrates and transforms to fine crystals of mullite, $3AlO_3{\cdot}2SiO_2$. The quartz dissolves slightly in the molten binder, but remains mostly as relatively large crystals that provide strength to the porcelain. As more feldspar is added to the porcelain, there is more liquid during firing and the ware becomes more translucent. As the amount of clay is increased, the powders are more easily formed and the resulting procelain becomes stronger and has lower electrical conductivity.

Dental porcelains must be relatively translucent and have easier forming requirements, so a composition with more feldspar and less clay and sand is usual. Porcelains therefore contain quartz and mullite crystals in a glassy potassium aluminosilicate matrix. The resulting properties are not sensitive to small changes in composition, forming methods, and firing temperatures, because as the firing temperature or amount of liquid is increased, more silicate and alumina dissolve in the liquid, increasing its viscosity and lowering its reactivity. Porcelains use inexpensive raw materials, are strong, chemically durable, good electrical insulators, and can be fired at relatively low temperatures and short times.

E. Alumina

Aluminum oxide is very hard (9 on the Mohr scale), stable to high temperatures, and chemically durable. Polycrystalline sintered alumina is very strong compared to most ceramics, although it is brittle. Its strength, abrasion resistance, and chemical inertness make it attractive for dental and bone implants.

Polycrystalline alumina is made by sintering the pressed powder, usually with polyvinyl alcohol or other organic material as a binder. The ease of sintering and properties of the dense alumina depend strongly on impurities in the starting material, and a wide range of quality and price is available commercially. Very pure alumina sinters only with difficulty at high temperatures (~1900°C). Addition of a small amount of magnesia gives a material with many fewer pores; commercial "Lucolox" is made this way and is translucent, so that it can be used as an arc tube in sodium vapor lamps. If the starting material contains a small amount of alkali (sodium or potassium) oxide, liquid forms during sintering, so that densification takes places more rapidly and at lower temperatures. However, the liquid forms a glassy phase at the grain boundaries, making the alumina weaker at high temperatures and less chemically durable.

Single crystal alumina (sapphire) is grown from the melt by the techniques mentioned below. It has excellent strength and chemical durability, but is very expensive.

F. Calcium Phosphates

Polycrystalline hydroxyapatite can be sintered to a strong, dense solid from powder prepared as described above.[10] The strength and opacity of the final product depend strongly on the phase purity of the starting crystals and the time and temperature of sintering. It is essential to start with crystals that all have the chemical composition of hydroxyapatite. This is achieved by stirring the precipitated crystals in the supernatant liquid for 24 hr at room

FIGURE 2. Transmission electron micrograph of ion-thinned hydroxyapatite fired at 1100°C for 1 hr. (Magnification × 75,000.)

temperature or shorter times at higher temperature. If these crystals are fired damp at 1100°C for 1 hr, the solid has theoretical density and few or no pores (Figure 2) and is translucent. If the sintering is carried out above about 1200°C, the hydroxyapatite partially decomposes and forms second phase particles on the grain boundaries, reducing strength and increasing opacity. If the starting material is not 100% apatite, the sintered product contains α or β whitlockite, $Ca_3(PO_4)_2$, and is again weaker and more opaque.

Hydroxyapatite shows great promise as a bone and tooth implant material, because it shows no foreign body reaction and bonds directly and strongly to bones without any intervening fibrous layer.[11] Whitlockite can be made into a dense, strong material by the same process as hydroxyapatite[12] and may find use as a bioresorbable implant.

G. Calcium Aluminate

Porous materials have been suggested as possible for implants; if the pore size is greater than about 50 μm, bone grows into the pores. However, it is difficult to get complete growth into the pores, and residual porosity may be a site for infection.

Porous calcium aluminate was made by sintering fine alumina powder with coarse calcium carbonate at 1450°C for 24 hr.[13] Different pore sizes resulted by using different sizes of the calcium carbonate particles; the pore size was about the same as the particle size. The calcium decomposes to calcium oxide, which then reacts with alumina to form two calcium aluminate phases, $CaO{\cdot}Al_2O_3$ and $3CaO{\cdot}5Al_2O_3$. With a starting composition of about 27% CaO and 73% Al_2O_3 (by weight), the amounts of the two phases were about those expected at equilibrium (20% $CaO{\cdot}Al_2O_3$, 80% $3CaO{\cdot}5Al_2O_3$). Porous $CaO{\cdot}TiO_2$ and $CaO{\cdot}ZrO_2$ were also made by reaction of calcium carbonate with TiO_2 or ZrO_2.

H. Hot Pressing

For certain applications of high value and limited sample numbers, various special techniques of ceramic manufacture are useful. Materials that are difficult or impossible to sinter to dense enough bodies can sometimes be sintered by a combination of heat and pressure called hot pressing. The main problem in hot pressing is to find satisfactory materials for dies that will withstand pressures of 30 MPa or higher at temperatures above 1000°C. Graphite is most widely used, but is quite expensive and has short life as a die, so hot pressing is only used for pieces of high value.

Powders with large crystal size can often be sintered by hot pressing even if they would not sinter without application of pressure. Hot pressing is slow because of time required to heat and cool the dies for each piece.

I. Crystal Growth

Sintering of powders is used to make ceramics because their melting temperatures are usually too high for molding from the liquid. However, for certain special purposes growth of crystals from the melt can be used to make ceramics. Again this is a high-cost process because of crucible problems and difficulties in producing large numbers of samples. Thus, crystal growth is usually used to make single crystals of ceramics for high-value applications. In the Czochralski method, a crystal is pulled slowly from a melt; this technique is used to make alumina (sapphire, corundum, ruby), garnets, and other valuable crystals. In the Verneuil process, powder is fed through a flame into a liquid pool on the crystal surface.

J. Vapor Processes

Thin films of ceramics with special chemical, electrical, magnetic, or optical properties can be grown by evaporation and deposition from the vapor. In addition, chemical reaction of gases is used to form ceramic solids (chemical vapor deposition, CVD). Boron and silicon carbide can be made by reducing BCl_3 or $SiCl_4$ in a hot vapor, with subsequent deposition onto a substrate of the form desired, e.g., a fine tungsten wire for fibers. Silica can be made by oxidizing $SiCl_4$.

IV. SINTERING

When a sample is fired, the individual crystals in the powders are joined together forming a solid, porous mass; as firing continues the pores are removed, finally forming a compact, dense material. Sintering can occur from entirely solid powders or from a mixture of solid particles and liquid; the latter process is sometimes called vitrification.

A. Solid-State Sintering

The driving force for joining powder particles into a dense solid is reduction of surface energy. Three stages in solid-state sintering are usually recognized: initial, intermediate, and final. In practice these various stages overlap and are not always clearly defined, but they provide a useful basis for classification and a beginning at understanding.

In the initial stage, individual particles join together by growth of necks between them. The usual mechanism of transport in this stage is bulk or volume diffusion. Material is transported from the interior of the particles to the neck region, and the total volume of the powder compact decreases. A grain boundary forms between adjoining particles, and diffusion may be more rapid in this boundary. The particles can also grow in this stage to reduce the surface area. This ''competitive'' growth involves shrinkage of smaller particles and growth of larger particles.

The neck region between particles can also grow by surface transport or by vapor transport if the vapor pressure of the material is high enough. In these cases the powder compact does not decrease in volume, so there is no macroscopic shrinkage even though the internal surface area is reduced.

The amount of volume shrinkage in the initial stages is proportional to some fractional power (about one half) of the volume diffusion coefficient and the solid-vapor surface energy of the material. It is also approximately proportional to the square root of time and inversely proportional to the particle size. In practice it is difficult to change the sintering time or the surface energy very much. Thus, the main parameters that can be changed to influence the rate of sintering are temperature and particle size. The diffusion coefficient is exponentially dependent on temperature, so increasing the sintering temperature can strongly increase the rate of sintering. Decreasing the particle size strongly increases the rate of shrinkage; ''active'' powders are those with very fine particle size.

When materials are sintered by hot pressing, the external pressure packs particles more tightly and increases contact area. Individual crystals can also be plastically deformed; there is disagreement as to whether or not this deformation aids sintering, other than packing particles more tightly.

In the intermediate stage of sintering, a continuous matrix of solid particles is joined together, leaving an interwoven network of continuous pores. The grain size of the solid matrix grows during this stage to reduce total grain boundary area, and the internal surface area decreases as pores are removed, causing the bulk volume to shrink. Volume diffusion is the probable mechanism of material transport during this stage, enhanced by more rapid diffusion in the grain boundaries. If the diffusion is controlled by a vacancy mechanism, it is usual to imagine shrinkage occurring by transport of boundaries and hence to the sample surface.

Most models of the intermediate stage assume that the continuous pores lie on grain boundaries. Thus, as the grain size increases the pore fraction decreases. Pores should stay with grain boundaries only if the pores are very small and consequently quite mobile or if the grain boundaries move very slowly. In many circumstances the boundaries will break away from the pores, leaving the pores isolated in a grain. On the other hand, the pores can pin the boundaries, slowing their motion. The cylindrical pores become spherical as the boundaries break away from them, leading to the final stage of sintering.

In the intermediate stage of sintering, the rate of shrinkage is again strongly dependent on the temperature through the volume diffusion coefficient and also depends on the rate of grain growth. Grain growth is also directly related to temperature, because it requires atomic rearrangements. The driving force for grain growth is the grain boundary energy, so the rate of grain growth is higher for smaller average grain size.

In the final stage of sintering pores are isolated in a continuous, dense matrix of material. In some compacts this stage may not be reached, and the material remains porous.

The position of the pores with respect to grain boundaries is a controlling factor in the rate of final stage sintering. Material (or vacancies, or gas in the pore) diffuses rapidly in the grain boundaries, so if a pore is on a boundary, it shrinks as vacancies or gases are transported to the material surface. However, if the pore is not near a boundary, its rate of shrinkage is much lower and in many practical situations is effectively zero, so pores away from grain boundaries remain in the material even for long sintering times.

Grain growth is therefore an important factor in final stage sintering. If the rate of grain growth is rapid compared to pore shrinkage, many pores will become isolated away from boundaries, and the final density of the material will be lower than theoretical. Alternatively if the rate of grain growth is slowed, e.g., by absorption of an impurity, the pores have time to shrink before the boundary migrates away.

Exaggerated grain growth (also called secondary recrystallization) involves the rapid growth of a few large grains in a fine grain matrix and can lead to trapping of many pores in the large grains. Thus, in many ceramic systems it is essential to attenuate exaggerated grain growth if one wishes a high density product. In alumina addition of a small amount of magnesia is found to reduce the rate of grain growth and especially exaggerated grain growth and leads to a translucent product (Lucalox) that can be used as an arc tube in high-efficiency sodium vapor lamps.

Removal of all pores in a sintered sample is rare; an example is shown in Figure 2 for hydroxyapatite.[10] The very small and uniform initial particle size, shown in Figure 1, is an important factor in achieving a pore-free material, even though the average grain size has increased from about 0.05 to about 0.6 μm during sintering.

B. Vitrification

If a liquid phase is present during sintering the process is quite different than for solid-state sintering. The liquid phase coats solid particles, helping to join them together. Material transport is more rapid in the liquid than in a solid. Thus, any pores or entrapped gases are removed more readily than in solid-state sintering. If the viscosity of the liquid is low enough, bubbles in it can rise to the surface. The solid particles can dissolve either partially or completely, which often raises the viscosity of the liquid.

Temperature is the most important parameter in vitrification. A minimum temperature (the melting temperature of a component) must be exceeded for the process to take place, and the viscosity of the liquid, which is strongly decreased at higher temperature, is an important parameter. The liquid must be fluid enough to coat the particles, and material transport in the liquid usually depends strongly on the temperature. Diffusion of gas in the liquid, which controls bubble shrinkage, is also strongly dependent on temperature.

The space between particles acts as a capillary, which induces particle wetting and coalescence. Smaller particles lead to a larger capillary pressure, enhancing wetting. A variant of vitrification is "reactive sintering", in which a solid phase dissolves in a liquid and then reprecipitates.

V. GLASS MELTING AND FORMING

A. Glass Compositions

The compositions, uses, and some properties of certain glasses are given in Table 1. By far the largest tonnage of glass is soda-lime. It is relatively easy to melt, uses cheap, readily available raw materials, and is easy to work and chemically durable. Its high coefficient of thermal expansion can lead to thermal shock and failure. Pyrex borosilicate glass has a lower coefficient of thermal expansion and is therefore less subject to thermal shock, so it is useful for cookware, labware, and telescope mirrors. It is also more chemically durable than soda-lime glass in neutral aqueous solutions, but not in strong acids or bases.

A large variety of special glass compositions have been developed for particular purposes. For example, high lead glass is used for "crystal" glassware because of its high refractive index and resultant "sparkle". Vitreous silica and the related Vycor glass are valuable for

Table 1
PROPERTIES, COMPOSITIONS, AND USES OF IMPORTANT SILICA GLASSES

Glass type	Corning number	Kimble number	Approximate composition (wt %) SiO_2	B_2O_3	Al_2O_3	CaO	MgO	Na_2O	Relative thermal expansion 10^{-7}/°C	Annealing temp (°C)	Softening temp (°C)	Uses
Soda-lime	0080	R-6	72.6	0.8	1.7	4.6	3.6	15.2	92	510	700	Windows, lamps
Pyrex borosilicate	7740	KG-33	81	13	2	—	—	4	33	555	820	Headlamps, cooking and labware
Soda borosilicate	7050	K-705	68	24	1	—	—	7	46	500	700	Sealing
Alkaline earth Aluminosilicate	1720	EZ-1	64	4.5	10.4	8.9	10.2	1.3	42	710	910	High temperatures
Bioglass®		—	45	(6% P_2O_5)	—	24.5	—	24.5	—	—	—	Implants
Vitreous silica	7940	—	100	—	—	—	—	—	5	1050	1580	Crucibles, optics
Vycor	7900	—	96	4	—	—	—	—	8	910	1500	Crucibles, optics

high temperature stability, chemical durability, low thermal shock, and high optical transparency. However, vitreous silica must be melted at temperatures above 1740°C and so is quite expensive. Vycor glass is made by a special lower temperature process described below.

The bioglass composition in the table has been found to bind strongly to bone.[14] It was first thought that this binding resulted from the calcium and phosphorous in the glass; however, Walker has shown that a variety of glass compositions will bind to bone whether or not they contain these elements.[15] Walker proposes that the requirement for bonding is fine porosity of a size of roughly 10 nm in the glass surface. Thirsty glass, which is a porous glass of the Vycor composition, also bonds to bone.[15]

B. Melting and Homogenization of Glass

By far the most important factor in the homogeneity of glasses is the care with which the raw materials are mixed. For the most homogeneous glass, the particle size of the starting powders should be very small, 1 μm or less. The powders can be mixed vigorously in an organic liquid such as ethanol, e.g., in a blender. Then the alcohol is removed by heating at ~80°C. This method is much more effective than crushing and remelting the glass.

The raw materials for glasses are quartz sand (SiO_2) sodium carbonate (Na_2CO_3), calcium carbonate (CaO), and sodium borate ($Na_2B_4O_7 \cdot 10H_2O$). The source of alumina can be a mineral such as feldspar. As the batch mixture is heated, adsorbed water and water of hydration are first driven off; then at about 700 to 800°C the carbonates begin to decompose and alkalis melt. At higher temperatures the refractory solids dissolve in the liquid alkali. A typical melting temperature for soda-lime glass is about 1300°C; Pyrex borosilicate requires a somewhat higher melting temperature.

Melts can be stirred, but careful mixing is more effective for homogenization. Usual batch melting times are a few hours. During this time some alkali oxide vaporizes, so the time should only be long enough to ensure complete melting.

The viscosity of a glass is strongly dependent on temperature and is the most important property that influences melting conditions. The viscosities of some glass are shown as a function of temperature in Figure 3, together with various temperature points.

C. Fining

The most difficult step in melting glass is removal of bubbles (fining). Bubbles bigger than about 1 mm in diameter rise to the melt surface in a reasonable time at glass melting temperatures (viscosity ~100 P), but bubbles smaller than about 0.5 mm rise slowly and are only removed by diffusion of the gas in them into the glass melt.

A number of empirical methods of fining glass have been developed over the centuries; the mechanism of these methods is still uncertain. Lowering the melting temperature ~100°C for a time (up to 24 hr) after the glass is thoroughly melted seems to help to remove bubbles. In commercial practice, a number of chemical additives are found to aid in bubble removal. The most common of these are arsenic or antimony oxides and sulfates, added to about 0.5 wt %. With arsenic and antimony, it is helpful if some of the soda is added as the nitrate instead of the carbonate.

In commercial practice, glass is melted in large tanks,[4] some up to 100 ft long. The raw materials are fed into one end of these rectangular furnaces, and the molten glass passes from this end through a hole in a refractory wall into the refining portion and then out of the tank to be formed into desired shapes. These furnaces are heated by gas or oil flames across the top surface of the glass. These tank furnaces can produce more than 300 metric tons of glass a day.

D. Vitreous Silica and Vycor Glass

Vitreous silica must be melted at temperatures above 1740°C, the melting temperature of

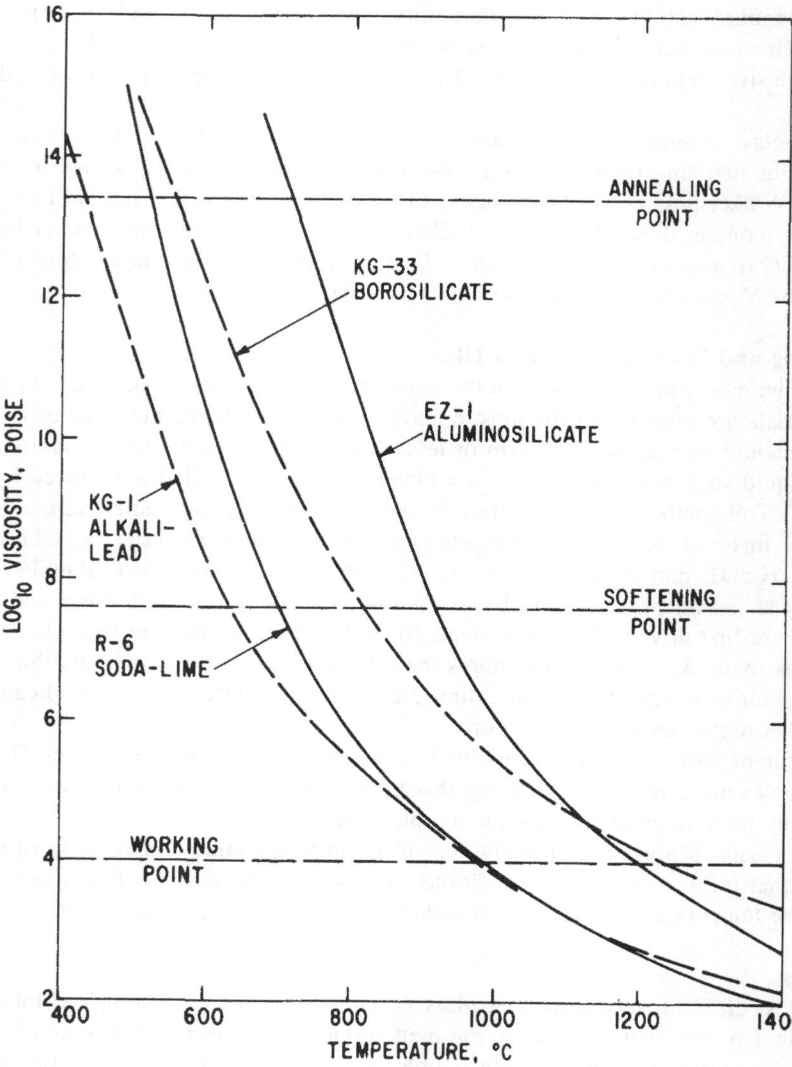

FIGURE 3. Viscosities of some commercial glasses listed in Table 1.

cristobalite, and therefore requires special materials and processing. In one method quartz chunks or sand are melted in cylindrical graphite crucibles in an inert atmosphere such as nitrogen. The crucibles are usually destroyed by the melting, making the process quite expensive. Then rod or tubing can be drawn from the ingots, or large pieces can be cut directly from the ingots.

High-purity vitreous silica can be made by oxidation or hydrolysis of $SiCl_4$ vapor. In one method the $SiCl_4$ is mixed with oxygen and natural gas, fed through a burner, and deposited on a substrate or preheated sand or on a rotating air-cooled mandrel of aluminum. A plasma torch can also be used. This method gives either a high water (SiOH) content or a high chlorine impurity level, depending upon oxidation conditions.

Crucibles for melting semiconductors are made directly from quartz sand by arc-melting; the sand acts as its own container. Fibers of silica can be pulled from a preformed rod or other convenient shape. Vycor glass is made by etching out a phase-separated borosilicate glass to give finely porous ‘‘thirsty glass’’ and then heating the porous material at about 1000°C to form a dense clear glass.

E. Forming

Flat glass for windows was traditionally formed by drawing a sheet of glass onto a metal table and then grinding and polishing both sides. This process was expensive and time consuming. Recently the float glass process has replaced older methods. In this process the molten glass is drawn onto a bath of liquid metal, usually tin. This method gives smooth surfaces so that no grinding and polishing is needed. The two sides of the glass have somewhat different properties, becuase some tin is incorporated into the face in contact with the liquid tin. Tubing, rod, and fibers are drawn from ingots or directly from a tank furnace through dies.

Many complex shapes can be made by pressing the glass in a mold; however, for rapid and continuous forming of bottles and other containers a two-step blowing process is used. In the ''individual section'' machine, a gob of glass is first fed into a blank mold of metal. Then the neck is formed with an iron plunger, and the glass is blown into the blank mold. The blank shape is then transferred to the blow mold, which has the final shape of the bottle, and the glass is blown into this mold. The finished bottle is then ejected from the mold. Special pieces such as artistic ware and laboratory equipment are still blown by hand.

VI. GLASS CERAMICS

Glass ceramics are glasses in which fine, uniform crystals are grown by controlled heat treatment. They have higher strength, chemical durability, and electrical resistivity than the parent glass and can be made with low thermal expansion, giving excellent resistance to thermal shock. Important applications of glass ceramics are as cookware, stove tops, radar domes, electrical insulators, and telescope mirrors.

Glass ceramics are one of the few really new engineering materials developed in the last decades. To make a glass ceramic, an article is first formed into a desired shape by conventional methods of fabricating glass (see above), and then it is heated first to nucleate and then grow the crystals throughout the glass. The resultant composite structure of fine crystals held together in a glassy matrix has improved properties.

Glass ceramics that contain a high density of fine crystals are white and opaque, resembling porcelain, because of light scattered by the crystals. However, it is also possible to make glass ceramics that are translucent or even transparent, so they can be used as optical materials such as telescope mirrors; a low thermal expansion is also advantageous for this application. Opal glasses, traditionally used for decorative purposes, have the same microstructure of fine crystals in a glass matrix as translucent glass ceramics.

Two classes of glass compositions give the best glass ceramics, the lithium aluminum and the magnesium aluminum silicates. The crystalline phase in the former can be β-spodumeme ($2Li_2O{\cdot}Al_2O_3{\cdot}4SiO_2$), β-eucryptite ($2LiO_2{\cdot}Al_2O_3{\cdot}6SiO_2$), β-quartz or lithium disilicate ($Li_2O{\cdot}2SiO_2$), depending upon the base composition. Cordierite ($2MgO{\cdot}2Al_2O_3{\cdot}5SiO_2$) crystals grow from the second type.

To initiate or ''nucleate'' uniformly the fine crystals in a glass ceramic, a special nucleating agent is added to the basic glass composition. Titania (TiO_2) or phosphate (P_2O_5) are effective nucleating agents in many glass ceramics. Fluorides are used as nucleating agents in aluminosilicate glasses and in soda-lime glasses, resulting from the traditional application of fluorides in opal glasses and glazes.

VII. CARBONS

A. Graphite

Carbon articles are formed by mixing coke filler particles with a binder such as coal or petroleum pitch. The coke itself is made from pitch by heating it to above 400°C, during which the aromatic hydrocarbons of intermediate molecular weight (~300 daltons) in the pitch are polymerized to the anistropic, rigid, solid coke particles. The mixture of coke in pitch is molded or extruded and heated in a packed bed to "set" the binder into the desired shape. Then the article is baked at 800 to 1000°C to pyrolyze the pitch binder into solid bridges that hold together the original coke particles. The result of baking is a porous carbon of low strength and electrical conductivity. To improve these properties, the article is heated in an electric furnace at a temperature from 2400 to 3000°C. This treatment orients the carbon atoms in the baked article into the hexagonal graphite lattice and orients the hexagonal layers. The result is a stronger, more conductive material called graphite.

B. Glassy Carbon

Glassy carbon is produced by pyrolyzing cross-linked organic polymers such as cellulose, phenolformaldehyde, and polyfurfuryl alcohol under carefully controlled conditions. The microstructure consists of random crystallites about 5 nm in size, in contrast to the ordered layer structure of graphite. The density of glassy carbon (1.4 to 1.5 g/cm^3) is lower than that of graphite, but glassy carbon has low porosity and permeability. The mechanical properties are similar to those of silicate glasses; the strength and modulus are higher than those of graphite. Glassy carbon is chemically inert compared to graphite, probably because of its low porosity. Glassy carbon is quite expensive to manufacture, which has limited its use to specialty applications.

REFERENCES

1. **Kingery, W. D.,** *Introduction to Ceramics,* John Wiley & Sons, New York, 1960.
2. **Norton, F. H.,** *Elements of Ceramics,* Addison-Wesley, Reading, Mass., 1974.
3. **Kingery, W. D., Bowen, K. K., and Uhlmann, D. R.,** *Introduction to Ceramics,* 2nd ed., John Wiley & Sons, New York, 1976.
4. **Douglas, R. W. and Frank, S.,** *A History of Glassmaking,* Foulis, G. T., Ed., Henley-on-Thames, England, 1972.
5. **Tooley, F. V., Ed.,** *Handbook of Glass Manufacture,* Books for Industry, New York, 1974.
6. **Doremus, R. H.,** *Glass Science,* John Wiley & Sons, New York, 1973.
7. **McMillan, P. W.,** *Glass Ceramics,* 2nd ed., Academic Press, New York, 1979.
8. *Encyclopedia of Chemical Technology,* Vol. 4, John Wiley & Sons, New York, 556ff.
9. **Heimke, G., Griss, P., Jentschura, G., and Werner, E.,** Direct anchorage of Al_2O_3-ceramic hip implants, *J. Biomed. Mater. Res.,* 12, 57, 1978.
10. **Jarcho, M., Bolen, C. H., Thomas, M. B., Bobick, J., Kay, J. F., and Doremus, R. H.,** Hydroxlapatite synthesis and characterization in dense polycrystalline form, *J. Mater. Sci.,* 11, 2027, 1976.
11. **Jarcho, M., Kay, J. F., Gumaer, K. I., Doremus, R. H., and Drobeck, H. P.,** Tissue, cellular and subcellular events at a bone-ceramic hydroxylapatite interface, *J. Bioeng.* , 1, 79, 1977.
12. **Jarcho, M., Salsbury, R. L., Thomas, M. B., and Doremus, R. H.,** Synthesis and fabrication of β-tricalcium phosphate (Whitlockite) ceramics for potential prosthetic applications, *J. Mater. Sci.,* 14, 142, 1979.
13. **Hulbert, S. F., Klawitter, J. J., and Leonard, R. B.,** Compatibility of bioceramics with the physiological environment, in *Ceramics in Severe Environments,* Kriegel, W. W. and Palmour, N., Eds., Plenum Press, New York, 1971, 417.
14. **Hench, L. L., et al.,** Development of bioglass, *J. Biomed. Mater. Res.,* 2, 117, 1971.
15. **Walker, M. M.,** An Investigation into the Bonding Mechanisms of Bioglass, M.S. thesis, University of Florida, Gainesville, 1977.

INDEX

A

B

C

I

J

K

L

M

N

For Product Safety Concerns and Information please contact our EU representative GPSR@taylorandfrancis.com Taylor & Francis Verlag GmbH, Kaufingerstraße 24, 80331 München, Germany